AF574461

"THE SCENIC ROUTE OF PUGET SOUND"

PUGET SOUND ELECTRIC RAILWAY

THE QUICKEST ROUTE BETWEEN

SEATTLE AND TACOMA

MT. TACOMA OR RAINIER

TRAINS EACH WAY EVERY HOUR

(ONE HOUR & THIRTY MINUTES)
TIME CONSUMED

2 LIMITED TRAINS

(ONE HOUR & FIFTEEN MINUTES)
DAILY BETWEEN
SEATTLE & TACOMA

J. FURTH, PRES.
SEATTLE

STONE & WEBSTER
GENL. MGRS. BOSTON

W. S. DIMMOCK, MGR.
TACOMA

C J. FRANKLIN, SUPT.
TACOMA

A. H. MACKAY, COML. AGT
TACOMA

TICKET OFFICES
OCCIDENTAL & YESLER
FREIGHT- FIRST & MASS. AVE
SEATTLE
EIGHTH & A STREETS TACOMA

PSE No.516 sits at the Tacoma depot in 1923 while an Olympia bound connecting stage waits for passengers. To the left of No.516 is one of Tacoma Railway & Power's Stone & Webster 200 series cars used on McKinley Avenue. (Boland photo, courtesy of Tacoma Public Library)

To Tacoma By Trolley

The Puget Sound Electric Railway

Warren W. Wing

Pacific Fast Mail Edmonds Washington

TO TACOMA BY TROLLEY
The Puget Sound Electric Railway

Edmonds, Washington

Library of Congress No.94-73942
ISBN No.0-915713-29-2

Book Design: Mike Pearsall
Maps, Wayne Hom

Other books by Warren Wing:
A Northwest Rail Pictorial Vol.I
A Northwest Rail Pictorial Vol.II
To Seattle by Trolley

Printed in Hong Kong

Below, PSE 500, 509 and 515 northbound on the trestle at Renton Junction in 1905. The house and barn in the background are still in existence. (Harold Hill collection)

Table of Contents

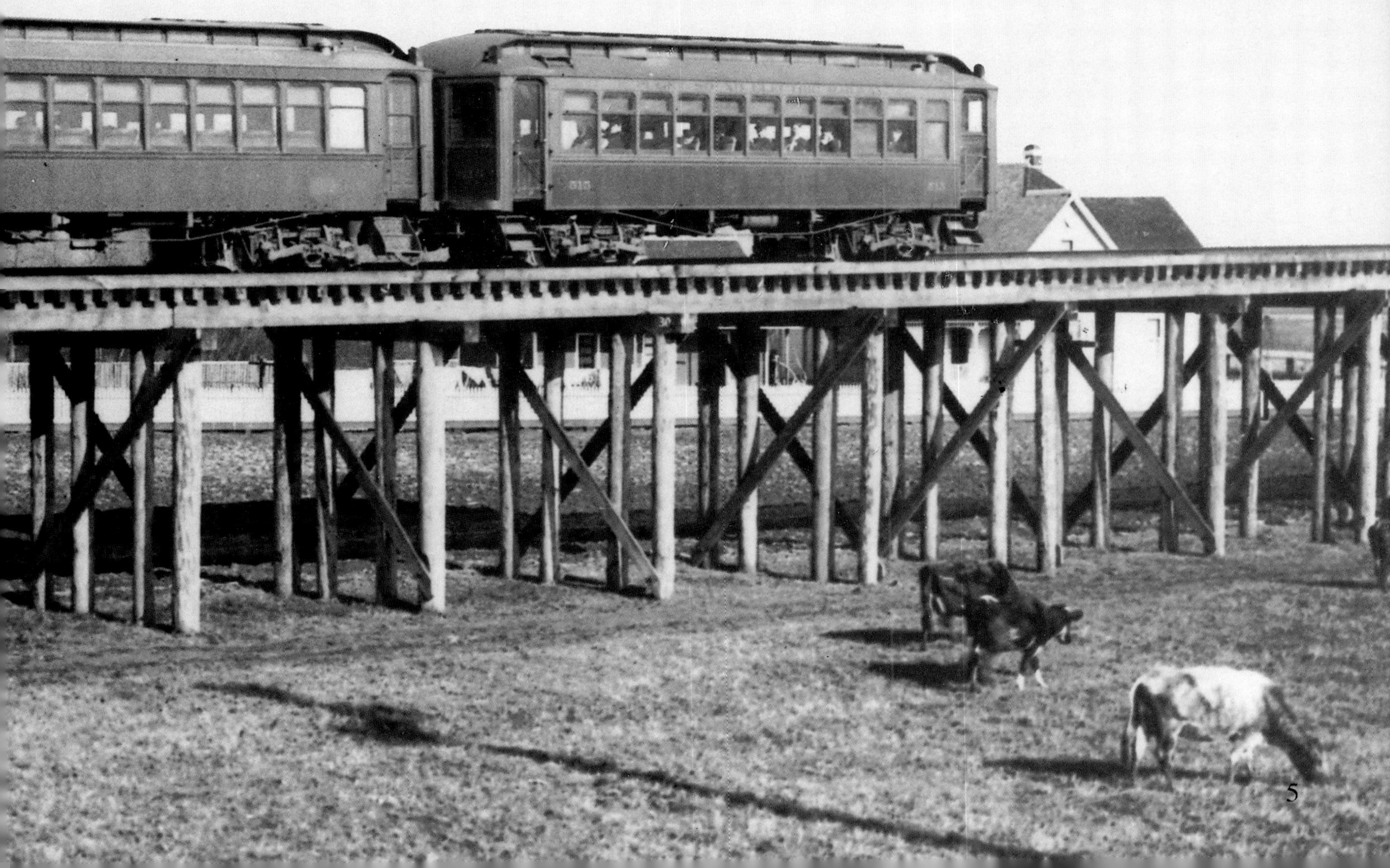

QUEEN
-23
512

Acknowledgements

Many times I've been asked how long it takes to produce a book like this. Well, when I think about it, I would have to say nearly forty-five years. That is if I include the research and the pleasant associations I've had with people of similar interests.

It began when I was delivering mail on the north side of Green Lake in 1949. Frank McLellan, a longtime movie theatre projectionist, trolley buff and son of a Green Lake streetcar motorman, lived on my route on Sunnyside Avenue. Frank introduced me to two men who helped spark my interest, James Turner and Lawton Gowey. Turner was an avid railroad photographer during an era (the 1920's) when very few people took railroad photographs. After Mr. Turner passed away, I was fortunate enough to purchase his collection of negatives. His remarkable photographs were the basis for my first book, *A Northwest Rail Pictorial.*

From his electric railway negative collection, Lawton Gowey taught me the art of contact printing. Later, I acquired an enlarger and in turn I taught Lawton how to make enlargements from the very same negatives. I might add that nearly all of the darkroom work done in connection with this book, as well as my three previous books, was done by myself.

Through Lawton, I met a number of other rail fans including Harold Hill. Hill also had similar interests and like Gowey, had a large collection of electric railway negatives.

In my estimation these two men were the leading electric railway historians in the Northwest. They were deeply interested in the Puget Sound Electric and both could have written this book if they were with us today. Without their generosity and the help of Jean Gowey and Irene Hill, this book could not have been written. Both Irene and Jean permitted the use of their husbands' extensive photo collections.

I expressed my appreciation to Jean Gowey many times before her recent passing and to Irene Hill as well.

There are many others, who provided materials and information, whom I wish to thank. Namely:
Clinton Betz, a Renton native, who shared his memories of the PSE; Viola Baker, for a photo of her conductor husband on a work train at Willow Junction; Patricia Cosgrove, museum director at the White River Valley Historical Society Museum, in Auburn, Washington, where I did much of my research; Barbara Campbell, who furnished me a copy of the photo showing me doing the research; Dan Cozine, for the Northern Pacific joint facility photographs; Richard Engeman at the University of Washington special collection section of Suzallo Library, who produced the Arthur D. Kempster scrapbooks; Jim Graves making available his collection of PSE timetables; Andy Hansen, who reviewed the captions of my photographs; Stan Greene, at Renton Historical Society; Wayne Hom, who drew the

A practically brand new PSE No.512 with trail car loading at the Occidental Avenue depot in Seattle. The 512 was placed in service in 1907 and remained until the end. The exact year of the photo could probably be determined by the make and model of the automobile coming toward the photographer. The Stone & Webster insignia on the front of the interurban can be seen below the motorman. (Museum of History and Industry)

maps for "To Seattle By Trolley"and this book; Bill Janssen for the pretty PSE brochure; Daniel Kerlee, who loaned me a number of postcards from his collection; Elaine Miller at Washington State Historical Society in Tacoma; Rick Caldwell and Carolyn Marr at Museum of History and Industry; Wendy Morgan of Tukwila Historical Society; Virgil Napier, who worked at PSE's car shops in Kent in the late 1920's; Daniel W. Brady for the rare 1902 pass; Gary Reece at Tacoma Public Library; Emery Roberts for the photos he took of former PSE freight motors on the Tacoma Municipal Belt; Vernon Rutledge, who took photos at the Bay Street yards after the cessation of streetcar service in Tacoma and who most recently gave me his negative collection; and lastly John Taubeneck for furnishing me an inventory of PSE equipment that was for sale following abandonment. My thanks to my brother, Dan Lynch, for giving my wife and me one of his company's surplus computers. His generosity and my wife Pat's persistence surely made this project so much easier. My friend, Noel Holley, spent a quiet Sunday afternoon proof reading my captions and text. Noel is a railroad author in his own right, having written two volumes on the Milwaukee Road and one on the "Kite Route", an early day interurban, between Denver and Boulder, Colorado.

For all the above and the nice people at Pacific Fast Mail, Don Drew, Verna Rauscher and Mike Pearsall, with whom I've had such a pleasant association the past several years; thank you for all your help and cooperation. Needless to say, it couldn't have been done without you. If I have overlooked anyone, I am truly sorry, it was not intentional.

Warren W. Wing

Seattle, Washington, 1994

Opposite, car No.523, running as a Limited, flies past signal No.170 (17 miles from Tacoma) kicking up dust at almost 60 mph. This speed was not uncommon on this stretch of tangent track. The car is northbound between Auburn and Thomas not long after the signals were installed in 1914. (General Signal Company, Author's collection)

TACOMA
LIMITED
SEATTLE
523

Limited train No.14 is ready to depart from the Seattle terminal at Occidental Avenue and Yesler Way. This photograph dates from late 1902 after the Seattle-Tacoma Interurban was reorganized as the Puget Sound Electric Railway. (Harold Hill Collection)

1. THE BEGINNING

September 25, 1902 is a significant date in northwest transportation history. For that day marked the inaugural run of the Seattle-Tacoma Interurban[1] Railway. While not the earliest, (that distinction goes to the East Side Railway, a 15-mile line between Portland and Oregon City in 1893), the Seattle-Tacoma Interurban was the first true high-speed electric railway, built to main line standards, to operate in the region.

It is difficult for us today to imagine just how important the coming of the interurban was. At the turn of the century cities along Puget Sound were growing rapidly; yet transportation was often crude and slow. It depended on the horse and wagon over bad roads and the whims of the steam railroads who were not against charging high fares and exorbitant freight rates. Residents of the White River Valley between Seattle and Tacoma wanted something else. For years they dreamed of another connection to Seattle besides the Northern Pacific. During boom times after the great Seattle fire of 1889, they discussed numerous propositions, but the Panic of 1893 put a crimp in any new transportation venture.

Then in 1900 Fred E. Sander stepped in and while new to the Seattle area, he became involved in real estate and railroad promotion. He built the Yesler-Jackson cable railway and then the Grant Street narrow gauge streetcar line. This route started at 2nd and Yesler and went south for some distance over wood piling along what is today Airport Way to Georgetown.

In October 1900, Sander secured a franchise to build an interurban line to Tacoma. The route would be via Jackson Street, 4th Avenue, across the tide flats and then south through the White River Valley to Tacoma. The names, in the Articles of Incorporation, filed on January 15, 1901, included O. O. Calderhead, Harry G. Ballou, Thomas B. Carter, John B. Neagle, William P. Trimble, and Fred E. Sander. They planned to double track and standard gauge Sander's Grant Street Line.

Traveling to New York to enlist eastern capital, Sander wired that one million dollars of new funds had been assured. In February 1901, it was reported Sander had issued a contract for construction of the right of way between Seattle and Tacoma. Also, an application for construction through the Puyallup Indian Reservation was pending with the U.S. Interior Department. The new interurban would be 33 miles long and be operational by the end of 1901.

Sander had planned to extend the Grant Street Line to Tacoma but before this could be accomplished it went into receivership and was sold at foreclosure on July 10, 1901. W. G. Grambes, later with Seattle Electric, took charge. Under his management the line quickly began to pay off expenses. It was then sold to Dexter Horton and Company and later came

[1] The term "interurban" is from the latin meaning "between cities". It was coined by Indiana state senator Charles L. Henry who wanted a word to describe several intercity electric railways with which he was involved . Henry has been called the "father of the interurban".

under Seattle Electric control.

Earlier, in December 1900, a plan for another line to Tacoma was unveiled; a plan directly in competition with Sander's extension of his Grant Street Line. One of the promoters was Jacob Furth, a Seattle banker. Tacoma promotors Henry Bucey and John Collins were also among the original franchise holders for the new line.

About this time, the King County Fair Association secured 209 acres two and one-half miles south of Georgetown. As the new Seattle-Tacoma Interurban closely paralleled the Grant Street route, by April 1901 it was not certain which road would serve the new fairgrounds.

On May 5, 1901, Jacob Furth, President of both National Bank and the Seattle-Tacoma Interurban, announced the electric line would be built to the race track and that the new interurban would be operational by January 1902. He also announced,

> "We have purchased steel rails and have arranged for equipment. There will be plenty of power and the cars will be the best to be had. There will be a first class, thirty minute service between the two cities".

Local newspapers hailed the opening of the White River Valley to rapid transit. King County had granted the necessary franchises and the U.S. Interior Department had approved the right of way through the Puyallup Indian Reservation. Newspapers also stated the line would go south on First Avenue, then on private right of way along a county road, known as the White River Road (Airport Way), which extended about two

Around the turn of the century narrow gauge Grant Street Electric Railway car No.12 sits, at what is today, Airport Way and Adams Street in Seattle. This line had its terminal at 2nd and Yesler and went south along the shore line below Beacon Hill on trestle work to the vicinity of present day Spokane Street. It was standard gauged soon after construction of the S-TI. The Grant Street Railway was built by Fred Sander after he completed the Yesler Way cable car line. (Lawton Gowey Collection)

miles south beyond the county fairgrounds.

By mid-May, hundreds of men were at work between Georgetown and Van Asselt, laying track and getting the line ready for the September opening of the fairgrounds. Residents south of the city were excited over the activity and rapid progress. By May 17th, financing of two million dollars was made available by Old Colony Trust of Boston.

Originally the line was to go south along the west bank of the Duwamish River. Instead it followed the old Seattle & San Francisco Railroad right of way to Englewood on the Grant Street Line (Airport Way & Lucille St.) as far as the fairgrounds. Leaving the county road and fairgrounds, it planned to follow the east bank of the Duwamish River to the Squire property, where it would cross the river on a combination wood and iron bridge. From there it would follow the west bank of the Duwamish to a point nine miles south of Seattle, where it would cross the Green River and go south, straight through Kent and Auburn. Southeast of Auburn, it was to go up Stewart's Canyon (Jovita Blvd.) and over the hill to the Puyallup Indian Reservation.

The company built a pile bridge on 1st Avenue over the tide flats. This work was begun by Henry Bucey and associates before the line was purchased by Stone & Webster. First Avenue South had become a sand spit due to dredging in the late 1800s.

Main construction was under contract to Hale and Smith, railroad contractors from Portland, Oregon. This company had long experience in railroad construction in the northwest with their first important contract the grading of the Northern Pacific right of way over the Cascade Mountains in 1886.

The main contract called for grading and culverting twenty-eight miles south of the fairgrounds. From Auburn, there would be six miles of heavy construction, including a two hundred foot tunnel at Stewart's Point plus a number of deep cuts and fills.

The late rains of summer in 1901 proved beneficial to the contractor as it softened the soil for grading. This part of the contract was to be completed by November 1st and was to include fencing of the right of way.

The new interurban was creating quite a stir. Local residents wondered what would happen to the steam railroads and boat lines once the new line was operating. Local newspapers supplied the answers in high-minded style.

"They will go on just the same. They will do business of old".

"Will the Interurban pay for itself?"

The reply:

"In the development of the country through which it passes and in augmenting the population; in bringing into touch centers with markets and trade center now entirely without connection of any kind".

"The new factor in local transportation is not to entirely

Above, a typical view of the Northwest when old growth timber was king and the spotted owl unheard of. This view could easily be when the Seattle-Tacoma Interurban was still a logging road in south King County near Federal Way. (Author's collection) Right, S-TI Ry. steam engine No.2 is shown hauling logs to the mill at Milton prior to electrification. (Lawton Gowey collection)

revolutionize the business by stealing away the business enjoyed by other capital. It is to be a developer, hence the country's gain".

By late September 1901, Hale and Smith's main camp was at Stewart's Point (West Valley & Jovita Boulevard). Workers were busy grading out the side hill, where the work was the heaviest, while two gangs were working on the tunnel approaches at Stewart's Point.

Located near Edgewood, at the summit of the "hogback" (as the separation of Puget Sound and the White River Valley was called), the tunnel at Stewart's Point was to be two hundred feet above the valley and, on the Tacoma side, 150 feet above the gulch. The maximum grade over the "hogback" was two percent. The rest of the line was perfectly level.

Electrical current on the new interurban would be 500 volts, the same as used in Tacoma and Seattle. Three substations receiving 25,000 volts, to be transformed into 500 volts, would be used. The current in Seattle would be produced at the Post Street station. Standard overhead trolley wire and outside third rail would be used for current pickup.

Stone & Webster of Boston, who owned and managed traction systems in several cities and who organized Seattle's independent street railways into one efficient system (Seattle Electric), was now in control. Since they owned the streetcar lines in Tacoma as well as Seattle, the Interurban would be the connecting link.

By early October 1901, with track completed, steam engines were running up and down the valley on the new Interurban. Valley residents were suspicious since they thought this might be a Northern Pacific scheme to utilize the tracks for a new way into Seattle. While it was true that some poles were in place, citizens south of Seattle wouldn't believe otherwise until they saw electric cars running on the rails.

Work on the Interurban stopped abruptly on January 1, 1902. The contractor declared he was losing money and refused to continue. After several weeks, the bonding company made good to the contractor, and work resumed. Then on March 19, work was again halted when John Hale, the contractor, died of Typhoid Fever. Representatives of the bonding company again took charge and the work continued.

In spite of a defaulting subcontractor and bad weather, rails had been laid to within sight of Black River by mid-March. Steam locomotives from both ends of the line were hurrying to the "front" with loads of steel rails, spikes, bridge timbers and other railroad supplies. By April the tracks were in daily use by construction trains through the valley almost to Auburn. On what was known as the hill section, seven miles of track were down and construction trains were using the new tunnel.

By mid-April 1902, Seattle Electric crews were rebuilding the Grant Street Line to standard gauge and building a new drawbridge across the Duwamish River to South Park. Once the change-over was made the old narrow gauge cars were shipped to Tacoma where the streetcar lines were still narrow gauge.

Meanwhile in early June new interurban cars began arriving at Seattle Electric's Georgetown car barn. Here, they were fitted with trolley poles and third rail shoes. Nine cars were on line at Georgetown by the 13th and a trial run from Georgetown to Race Track (Meadows) was scheduled for the following week. The company announced general headquarters would be in Tacoma while the car repair shops would be in a new building in Kent, south of the Kent Depot. Fares to Tacoma would be sixty cents one way and one dollar for a round trip. They also announced that eleven cars were ready for service; five motors and six trailers with the motors to have smoking sections and space for light express. The trains would leave Seattle and Tacoma at one and one half hour intervals with three express trains daily, stopping at Kent and Auburn only, taking one hour and fifteen minutes. The accommodation trains would stop on signal at all stations and take one hour and thirty minutes.

Regular service to Meadows began on August 6, 1902 with motorman Pennington and conductor Scott in charge of Seattle Electric car 89. Pennington was formerly with the Rock Island Railroad while Scott was a former Ballard line motorman. The following day, steam was turned on for the first time at the Post Street power house.

The first test of the new third rail system took place on August 21, 1902, using car No. 500, from Van Asselt Station to Riverside and return. A week later a trial trip to Kent was made with the car attaining a speed of 55 mph along this straight stretch of track.

At the time of installation, the company claimed the exposed third rail was not dangerous to people and animals. If anyone came into contact with the rail they'd only receive a mild shock. However, after the trial run, while several men were working alongside the third rail near Kent, the power was accidentally turned on in Seattle causing every man close by to be knocked to the ground as if struck by lightning...so much for the "mild shock". In September, a laborer working on the new line was electrocuted when he accidentally fell across the track and third rail.

By late August all rolling stock had arrived at Georgetown except two cars damaged while in transit. They were returned to the Brill factory in St. Louis for repairs. Officials reported the accident would not delay the opening of service later in September.

Service began from 1st and Madison Streets in Seattle to Renton on September 14th. Stops were scheduled for Georgetown, Race Track, Riverside, Foster Siding, Black River and Renton Junction. A running time of forty minutes was set with hourly service from 6:25 AM until 4:00 PM. The fare was 15 cents each way. A small building on the southwest corner of 1st and Jackson was used as the main waiting room and ticket office.

On the same day, a party of company officials was taken over the entire line to Tacoma and return. The party consisted of Jacob Furth, President; George Dickenson, General Manager; E.P. Robinson, Assistant Manager; George Donworth, Secretary; J. P. Lutes, Supt. of Light and

No.508 and 509 are typical of the first series of cars used on the line. They arrived from the builder, J.C. Brill, in early June 1902. 508 was originally built as a trailer, later motorized while 509 was built as a motor coach. The 508 above, is shown at the Kent barn soon after arrival and 509 below, is posing at Tacoma station. This car was sold to Seattle City Light after abandonment and used for many years on City Light's popular Skagit tours. (Both: Author's collection)

These two photos were taken either before or soon after the line opened. The rather grainy view of the new right-of-way in Jovita Canyon is from an industry magazine STREET RAILWAY JOURNAL, which devoted an entire issue to the new interurban. (Harold Hill collection).The second photograph is at Milton. Note the newness of the ties, roadbed and substation. The man on the velocipede has just come down the 2% grade from Edgewood and whether he can peddle back up again is a matter of speculation. The Milton mill and PSE spur are on the left. (Renton Historical Society)

Power; Frank Bartlett, Chief Engineer; and L. E. Bradley, Superintendent.

Operation to Tacoma commenced on September 25th with cars leaving Tacoma at 6:15 AM and Seattle at 6:50 AM. In the beginning service was to be at two hour intervals. The route was 36-1/2 miles long with twenty-two stops in between. It was single tracked from Jackson Street in Seattle to Bay Street in Tacoma with a two mile passing track between Thomas and Auburn along with various sidings and short passing tracks at other locations. There were passenger shelters at every stop and depots at Kent and Auburn. Freight sheds were also located at the more important stops. Operating offices, along with quarters for the crews, were at Kent. The company carbarn was also located there, south of the depot.

On September 30th, service was cut back from 1st and Pike Streets in Seattle to a loop at 1st and Yesler due to problems with the cars. This prevented them from making the grade on 1st Avenue and it was two weeks before changes could be made. The stiff grade on 1st was the reason the Everett Interurban terminated uptown rather than making a connnection with the Tacoma Line at Yesler Way.

A high speed experimental trip, without passengers, was made on October 7th. Speed reached was seven miles in six minutes. This was the fastest ever attained by a train in the Northwest up to that date. Between Kent and Auburn the train made better than a mile per minute.

On the same day, a rate war began between the Interurban and the passenger boats plying between Tacoma and Seattle. The day the Interurban began, the little boats ran practically empty, with patrons deserting the ships for the faster electric cars. Since opening day of the Interurban, various ships, including the famed *Flyer*, suffered a 25 to 75% decline in business. The passenger boats slashed their fares to meet the competition but even this did not help; people crowded the Interurban even though there was not enough cars. In the long run, though, the boats won, outlasting the Interurban by two years. The regular two hour service provided by the Puget Sound Navigation's steamers, *Tacoma* and *Indianapolis*, ended on December 15, 1930.

On November 23, 1902 the Seattle-Tacoma Interurban became the Puget Sound Electric Railway. Incorporation of the new company was the following month and all properties of the Seattle-Tacoma Interurban were deeded to PSE by April 1903. Headquarters for Puget Sound Electric was moved from Kent to Tacoma. PSE was more closely allied to that city since it owned the Tacoma streetcar system (Tacoma Railway & Power Company) and held over 12,000 shares of its associated Pacific Traction Company.

On the front page of the *Seattle Post Intelligencer* for November 23rd, a photo appeared of an interurban car on one of the bridges between Seattle and Tacoma. The story with it read in part:

> "Novelty of the interurban has worn off. Traffic has settled down with some of the passengers returning to the passenger boats. Passenger business so far is up to expectations on the Interurban. While it is a great

PUGET SOUND ELECTRIC RAILWAY
SEATTLE - TACOMA INTERURBAN
PUGET SOUND ELECTRIC RAILWAY
MEADOWS TO TACOMA
KING COUNTY FAIR ASSOCIATION
1908 MEET
Admit One Lady
TACOMA TO MEADOWS
Admit One Gentleman
PUGET SOUND ELECTRIC RAILWAY.
ROUND TRIP TICKET.— GOING.
TACOMA to KENT
ROUND TRIP TICKET.— RETURNING.
KENT to TACOMA
"THIRD-RAIL SYSTEM"
Puget Sound Electric Railway
Pass I. A. Nadeau
GOOD UNTIL DECEMBER 31st 1902 UNLESS OTHERWISE ORDERED
No. 13
Pass DR. P. W. WILLIS
1903
Puget Sound Electric Ry.
RETURNING
SEATTLE
ARGO
GEORGETOWN
MEADOWS
SOUTHSIDE
FLORAVILLE
LOCAL TICKET
GOOD ONLY WHEN OFFICIALLY STAMPED ON BACK HEREOF AND FOR CONTINUOUS TRIP.
94219
Form L. B. 1
Puget Sound Electric Railway
1927
Pass
Mrs. Roy Kelly,
Wife, Trainman.
No. 442
BETWEEN SEATTLE AND TACOMA
GOING
SEATTLE TO TACOMA
48392
TACOMA TO SEATTLE
15748

convenience for the citizens of both Seattle and Tacoma, the Interurban means much more to the communities along the way. The day the first car ran, Kent and Auburn were lifted from the country-side right into the suburbs of two cities and fully appreciate what that means to them. It now means that Kent is as close to downtown Seattle as Ballard and Green Lake. Kent is now nestled alongside the Interurban like a bug in a rug. A Kent worker can now get to work in Seattle as quickly as workers from Seattle's northend. It has done more for Kent than any other community along the line.

"The real benefit will begin when trains are handling milk and small express. So far the management has been taken up with the passenger business. This part has gotten down to a smooth operation. In about ten days special express and freight trains will be handling milk into Seattle and Tacoma every morning and evening.

"Five trains are now in operation, handling all the business possible. More equipment will be added soon. The company policy at present is to get as much mileage out of each car as possible. Two more trains are not to be added until absolutely necessary. The midnight runs have been eliminated as there is not enough business to warrant their operation. Special trains will be operated to handle theatre parties or any considerable number of people wishing to remain in Seattle or Tacoma until a late hour, although arrangements must be made in advance.

"The towns along the line that were actually created by the Northern Pacific are surprisingly friendly to the new Interurban. The NP depot in Auburn has been moved closer to a wye so now it is about as close to downtown Auburn as the interurban depot".

The new interurban was on its way.

Car No.500 pauses near Orillia on what is probably a trial trip over the line prior to the beginning of regular service. (Harold Hill collection)

A scene at the Kent carbarn right after the start of service.. Car 509 is at left and boxcar N0.1551 and car No.508 are in the right background. City car No.126 is thought to be from the Seattle street railway system. (Clark of Kent, Author's collection).

A 1903 view of PSE No.506 and two car train northbound along Tukwila hill, north of Renton Junction. The line through here was double tracked several years later. This particular photograph was used on brochures to promote the interurban. (Harold Hill collection)

Above, looking like a scene from a western movie, Milton, Washington is shown during the early days of the interurban. At right is the depot and freight station and on the extreme right, the company owned mill. Today, Highway 99 and I-5 would be in the upper left hand corner. Two buildings, built during the 1920's ,are still standing along what is now Porter Way in old Milton. The business district has since moved to Meridian Road, a mile or so east. (Author's collection) Below, a southbound Tacoma train stops at Renton Junction around 1903. (Harold Hill collection)

324

The Author discovered this beautiful photograph in the Historical Society's file under ""Coca Cola". Here, motor car No.504 and two trailers are loading at the interurban's Seattle station on Occidental in 1903. (Washington State Historical Society)

Opposite, looking north on 1st Avenue South to Jackson Street in Seattle around 1905. PSE No.324, a Renton car is taking on passengers before heading south over the single track on 1st Avenue South. 1st Avenue was double tracked several years later when city cars began operating over 1st on their way to Alki and Fauntleroy.According to a patron on the Author's mail route, 1st South in those days was a sand spit as far south as Lander Street.. The small building on the southwest corner of 1st and Jackson is the Interurban's orignal depot. (Webster & Stevens photograph, Museum of History & Industry) Opposite above ,a closer view of car No.324 on Main Street in Renton. This car was another refugee from the Seattle Street Railway and lasted until the late 1930's. (Author's collection)

Another outstanding early photo at PSE's Seattle depot on Occidental. Motor car No.502 and the two trailers were the first cars on line out of Seattle on September 25, 1902 (Washington State Historical Society)

Opposite above, a three car train with motor car no.501 leading rounds the curve from Jackson onto Ist Avenue south in 1905. (Edwin Little, courtesy of Wilbur Little) Opposite, a two car train at the depot on Jackson Street. This photo was taken in late 1903 during the short time the terminal was located here.

CAPITAL BREWING C
506
505

The Interurban building at Occidental and Yesler in Seattle. A sign over the corner doorway reads "Puget Sound Electric Railway, Ticket Office and Waiting Rooms". Just behind the Yesler Way cable car is G.O. Guys drug store on the corner of 2nd and Yesler. Guy's was an early day chain store and was located there for many years. At right is PSE's parlor car No.527 on the rear of a three car train. (Museum of History & Industry)

Opposite above, PSE No.503 and 517 on Yesler Way in Pioneer Square enroute to the depot on Occidental. Behind the totem pole is the Merchants Cafe, still at the same location today. The Pioneer building on the left is also still there as is the old Interurban building beyond the Seattle hotel. The hotel has since been replaced by a parking garage. The building housing the ticket offices of the Burlington Route and Northern Pacific is also gone, replaced by a more modern structure. (Lawton Gowey collection) Opposite, looking north on 1st Avenue South to Washington Street in Seattle, circa 1905 (Author's collection)

HOTEL SEATTLE
NORTHERN PACIFIC EXPRESS CO
105
103
Burlington Route
NORTHERN PACIFIC
101

SILVER BOW HOTEL
MAGNOLIA BAR BOHEMIAN BEER
BAR WHISKY 5¢
BAR RAINIER BEER

An interurban train is about to turn into Occidental from Yesler Way. The Wild Rose wagon and team belonged to the Frye Packing Company while Augustine & Kyer was a long-time grocery company located in the Colman building on 1st Avenue at Marion Street. When the interurban first located their terminal at this busy site, it raised protests from local teamsters, yet the depot at Yesler and Occidental remained until the end. (Author's collection) At right, the energetic citizens of Allentown are shown building their own interurban passenger shelter. (Tukwila Historical Society)

At left, a group of passengers wait for the interurban at Tukwila in 1906. Below, train No.22 is speeding north near Tukwila station about 1910. From all indications the two car train is traveling in excess of 50 mph. (Both: Author's collection

Two views of the road crossing at Riverton. The top photograph was taken in 1905 by visiting GN telegraph operator, Edwin Little. The second taken a few years later when a second track was added. Riverton was the site of several serious accidents during the Interurban's lifetime. (Tukwila Historical Society)

Car No.500 on a special run opposite the company owned Renton coal mine. The mine opened about the same time as the interurban and shut down at the end of PSE operations in 1928. The tracks crossing the trolley line are those of the Northern Pacific, which also served the mine. Coal from the mine was hauled to the company's Post Street steam plant in Seattle and used to heat the downtown buildings. (Author's collection)

An overcast day view of a three car train stopped at Renton Junction station. The parlor car on the rear would date the photograph as early1907. In the background is the swing bridge over the Green River. The bridge was designed to open for ship traffic but was seldom used for that purpose after the Lake Washington ship canal was built. The lowering of Lake Washington reduced the Black River to a mere trickle. (Harold Hill collection)

PSE No.500 and two car train on Pacific Avenue in Tacoma. Note the dual gauge trackage; this is when Tacoma's street railways were still narrow gauge. The train is southbound en route to Seattle. (Harold Hill collection)

Opposite above, PSE's famous wood lined 182 foot tunnel was high above West Valley highway just north of the intersection of West Valley and Jovita Blvd. The tunnel remained open for a number of years after abandonment of the trolley line but was eventually dynamited shut. Both portals were recognizable until the 1950's but nothing remains today (Author's collection) Opposite, No.506 and three cars idling on the extra track at the 8th & A Street station in Tacoma. On the far side of Commencement Bay is Brown's Point and at left, just above the four story building, is the spire of the Northern Pacific headquarters building. From the newness of the cars it appears the photo was taken soon after the line began. (Washington State Historical Society)

In a wonderful photograph taken in the mid-teens, PSE Renton local No.125 is ready to leave Seattle for its daily 5:35PM trip. Renton trains made all stops between Seattle and Renton Junction. (Washington State Historical Society)

2. Early Success, First Troubles

That the new Interurban brought prosperity and growth goes without saying. Early on developers were selling home sites along the line from Riverton to Pacific City and many families were taking advantage and moving from the cities. One developer was advertising acre lots for $65, with $20 down and $5 per month. The interurban company was not only advertising their schedules in the weekly papers but the fact they furnished electricity for homes and businesses. One advertisement proclaimed,

"What's the use? Why worry over trimming lamps and using inflammable explosive gas when you can have 'ELECTRIC LIGHTS'".

In December 1907, the PSE operated their first new two-car train on a trial run. While it met with satisfactory results (according to the company) the public was more interested in the standard three-car trains with more seats than two-car trains with standing room only. About this same time, Seattle Electric was experimenting with city cars hauling trailers. However, riders preferred one car every five minutes, not two cars every ten minutes.

Operation of interurban trains within Seattle and Tacoma resulted in several strange arrangements with the street railway companies of each city. In Seattle, Puget Sound Electric paid Seattle Electric 5 cents per passenger carried over the city's streetcar rails. SE in turn paid PSE $3.25 for crew and equipment. Also, PSE paid the streetcar company $1.75 per passenger for power and use of tracks. In Tacoma a similar arrangement was in effect.

In May 1908, Pacific Traction in Tacoma planned to build a summer resort at the end of their new American Lake Line. However, the company felt the two saloons located there would be detrimental to patrons of the new resort. Unable to have the saloons removed, PT ripped up their tracks and moved everything a half mile away. Railway men declared it was the first time saloons had driven a railroad out of the neighborhood. At the new location, the trolley line purchased enough property to finally get rid of the saloons once and for all.

On September 26, 1908, Fred E. Sander sold his Ballard-Hall's Lake interurban line to Stone & Webster. Sander had visions of extending his Seattle-Everett Interurban Railway into Everett but it was S&W who eventually finished the line. S&W planned the Everett line to become another link in a traction "Empire" connecting Portland, Oregon with Vancouver, B.C.

Later that year, in December, Puget Sound Electric opened its new "Short Line" providing better service between Tacoma and Puyallup. The old line between these two cities had been operated by Tacoma Railway & Power for many years and in July 1909 it was purchased by PSE. In turn the interurban leased it back to TR&P for $9,000 per year. Puget Sound Electric also owned the Portland Avenue line (Puyallup Ave. to S.42nd St.) and leased it to TR&P.

On June 1, 1909, President William Howard Taft pushed a button at the White House in Washington, D.C. that opened the Alaska Yukon Exposition in Seattle. The interurban anticipated a dramatic increase in patronage since tourists were expected to flood the area to visit the fair. At the same time, Puget Sound Electric began advertising light and power to the valley citizens on a twenty-four hour basis.

In October 1909, Jacob Furth, President of Puget Sound Traction, Light and Power caused a stir by announcing fare increases. Furth claimed fares had been too low and were not bringing a return on their investment. He wanted new rates based on 2 cents per mile, the same as charged by main line railroads.

Naturally riders were up in arms. They did not believe Furth's claim and vehemently fought the fare increase. An editorial appearing in the weekly Auburn *Argus* took exception to the increase with this statement;

> "The interurban had more business than they could properly handle. The new rates are protested as burdensome and now the kind hearted corporation is taking this action to relieve the strain on their rolling stock and over-worked trainmen. The public along the line should assist this worthy cause. They can deal directly with the local merchants instead of running into the city. Yes, indeed, Gentlemen, make it five cents per mile and we will all stay home or walk".

Mass meetings were held at Duwamish, Riverton, Foster, Tukwila and Kent. At the latter, C.W. Horr, of Seattle, suggested the valley citizens consider annexing to the city in order to have control over the fares charged by the interurban. He claimed this was the surest way to get even with the company. Developer C.D. Hillman, spoke of his personal interest and said he would spend his last dollar, if necessary, to bring the company in line. He insisted Furth had promised low rates when he (Hillman) was buying land along the interurban.

To further fight the increases, negotiations were underway with the Milwaukee Road and the Northern Pacific to secure better local service and for handling all freight in and out of the valley.

In March 1910, the State Railway Commission ordered round trip fares restored to the rates in effect before October 17, 1909. The commission felt that round trip rates being double that of the single fares was "unjust, unreasonable and excessive". This order did not affect the single fare rates and applied only to points south of Edgewood, the Puyallup Short Line and the Renton Branch. A week later, PSE announced its intentions of appealing the Commission's order to the courts.

That fall, F. S. Pratt, Vice President of Stone & Webster, toured S&W's northwest properties, including PSE and TR&P. He announced street railways were declining and his company was not making promises that it could not keep. This statement was rather surprising since at the time street railways were at their zenith and faced very little competition.

Early in 1911, PSE announced their gross earnings for 1910 as $150,274 compared with $139,018 for 1909 while their net earnings were

Duwamish Valley residents were irate with the company's new fare increases in 1909. This situation went back and forth until 1913when new fares went into effect. To counter, local people tried everything from a sit down strike on some of the trolleys to sailing a tall masted ship upriver to force the company to open drawbridges that hadn't been open for years. The most effective method in fighting the fare increase however, was the beginning of the valley's own bus line shown in this photograph. (Tukwila Historical Society)

$44,237 for 1910 and $41,520 for 1909. The Company had 208 employees on the payroll with wages totaling $192,120.

By late January 1913, the Interurban was back in court fighting the State Railway Commission's decision regarding the fare increase of October 1909. The company argued that it was receiving only 1% on its investment when it should be more like 7%.

The *Seattle Star* ran an editorial against the increase. The newspaper claimed if Stone & Webster interests had erred in building the PSE, then it was their fault. The *Star* also asked,

> "Why doesn't the company offer the right of way, franchises and rolling stock for sale if the business doesn't pay? Why not offer the line at a fair price to someone who won't be burdened with watered stock upon which no dividends would have to be paid?"

The following year residents were still outraged over the fare increases, some going from 15 cents to 40 cents for single one-way fares. Tukwila claimed they were devastated, people were moving back to Seattle with businesses closing and property values dropping.

A PSE ad appeared in the May 4th edition of *The Seattle Daily Times,* explaining the company's reason for raising fares. It was the old story of Furth's two cents per mile claim and the poor return on their investments. Earlier, the *Seattle Star* had a cartoon showing a well dressed Seattle Electric official with money bulging from his pockets while selling pencils. He had a sign reading, "Help The Poor".

The same cartoon has a Duwamish resident standing nearby with a policeman. He has a sign labeled: "Public Service Commission", that read, "I tell you, he is faking, he has all kinds of money!"

It was reported later that George H. Nichols, one of the valley's leading developers, was known to have promises from Jacob Furth that there would be a five cent fare to Riverton when service began. Now the company was seeking an increase to 34 cents. They claimed new auto bus competition was causing heavy losses. Nichols went on to say that Puget Sound Traction Light and Power had made a profit of $1,589,000 in 1913, but for three years had purposely tried to show losses on the Interurban to justify a fare increase. Valley residents were making similar claims.

To protest, people came up with novel ways to fight the fare increase. One involved a group of local ladies who staged a sit down strike on one of the cars. Another was to harass the Interurban by leasing a tall ship in South Park and sailing it up river forcing the company to open draw bridges that had not been opened for years.

Yet with all the flap against PSE the company was correct in one regard; automobiles and buses, rolling on improved roadways, were taking riders away from the Interurban. While many did not realize it at the time, the slow ride toward the end had begun.

Below, a builder's photo of PSE's beautiful combine No.514. It was sold in 1929 and became a restaurant in Kent. It was later destroyed in a fire. (Harold Hill collection)

Above , a classic company photo of cars 510, 513 and 525 near the Kent depot, about 1910. In 1927, the 510 crashed into a truck south of Auburn rupturing the gasoline tank. A spark from the interurbans electrical system ignited a fire which soon spread to the cars wooden frame. Despite the efforts of the train crew, both truck and trolley were destroyed. Below, a rather poor reproduction from a traction magazine of Parlor car no.521 at Kent. Unfortunately no photos have surfaced showing this car in service. In this view the car is not motorized. The observation platform in the rear was open during the summer and closed in winter. There was an extra charge of 25 cents to ride it. This car was later rebuilt with a baggage and smoking section and renumbered 516. (Both: Lawton Gowey collection)

Here is combine 514 in service with a train leaving Seattle station on Occidental Avenue about 1910 (Roger Dudley photo, Author's collection)

No. 523, running as a late afternoon limited, is about to enter the single track at Black River. Here, it will cross the Green River without slowing as the conductor grabs the order hoop "on the fly". The track was on pilings at this point as the Duwamish River (the name change here) came very close to the hillside. Presently, the northbound lanes of Interurban Avenue are now where the limited stands. This scene dates from 1910 when No.523 had an end door. They were removed and the end closed several years later. (Lawton Gowey collection) At left, an interior view of 523 taken at the Seattle terminal. This car and four others in its class were open ended in summer and closed during the winter. There was an extra charge of 25 cents to ride these splendid cars. Originally built as non-motored trailers they were later motorized and converted to straight coaches. (Author's collection)

Looking down on James Street and Yesler Way in Seattle at the new 42 story L.C. Smith tower. James Street and Yesler cable cars are in view as is a city AYP car from Alki. Behind it is a PSE 520 class interurban. (Author's collection)

It was a rainy day in 1908 when the first PSE car arrived in Puyallup over the Short Line. This Jewett built trolley was one of seven similar cars rebuilt for the Renton and Puyallup branches (Daniel Kerlee) Below, an early view of Puyallup looking south on Meridian Street. The narrow gauge tracks belong to the Tacoma Railway & Power Company, a subsidiary of PSE. The TR&P line went south past the fairgrounds, then west up the hill, through Summit and Midland to a connection with the Spanaway line at Parkland.(Author's collection)

PUYALLU

WESTWARD																		First Class								Ti
Daily 247	Daily 245	Daily 243	Daily 241	Daily 239	Daily Ex Sun 237	Daily 235	Daily Ex Sun 233	Daily 231	Daily 229	Daily 227	Daily 225	Daily 223	Daily 221	Daily 219	Daily 217	Daily 215	Daily Ex Sun 213	Daily 211	Daily Ex Sun 209	Daily 207	Daily Ex Sun 205	Daily 203	Daily Ex Sun 201	Car Capacity	Miles From Puyallup	
M 244 246 L 11.00 PM	M 242 244 L 10.00 PM	M 238 240–242 L 9.00 PM	M 234 236–238 240 L 8.00 PM	M 232 234–236 238 L 7.10 PM	M 230 232–234 236 L 6.40 PM	M 230 232–234 L 6.05 PM	M 228 230–232 L 5.40 PM	M 228 230 L 5.05 PM	M 226 228 L 4.10 PM	M 224 226 L 3.10 PM	M 222 224 L 2.10 PM	M 220 222 L 1.10 PM	M 218 220 L 12.10 PM	M 214 216–218 L 11.10 AM	M 210 212–214 216 L 10.10 AM	M 208 210–212 214 L 9.10 AM	M 206 208–210 212 L 8.35 AM	M 204 206–208 210 L 8.05 AM	M 204 206–208 L 7.35 AM	M 202 204 206 L 7.05 AM	M 202 204 L 6.35 AM	M 202 L 6.05 AM	M 202 L 5.35 AM	WYE	0	PY Rainie
11.03	10.04	9.04	8.04	7.13	6.43	6.09	5.43	5.09	4.13	3.13	2.13	1.13	12.13	11.13	10.13	9.13	8.38	8.08	7.38	7.08	6.38	6.08	5.39	3	0.78	
11.07	10.09	9.09	8.09	7.18	6.48	6.16	5.48	5.16	4.17	3.17	2.17	1.17	12.17	11.17	10.17	9.17	8.44	8.14	7.44	7.14	6.44	6.14	5.42	4	2.43	
11.09	10.11	9.11	8.11	7.21	M 238 6.50	M 236 6.20	M 234 5.50	M 232 5.20	4.20	3.20	2.20	1.20	12.20	11.20	10.20	9.20	M 214 8.48	M 212 8.18	M 210 7.48	M 208 7.18	M 206 6.48	M 204 6.18	5.45	7	3.56	
11.12	10.14	9.14	8.14	7.24	6.53	6.23	5.53	5.23	4.22	3.22	2.22	1.22	12.22	11.22	10.22	9.22	8.51	8.21	7.51	7.21	6.51	6.21	5.47	7	4.70	
M 28 68 11.15	M 28 68 10.18	M 26 66–28 9.18	M 24 26–66 8.18	M 240 64–22–24 7.28	M 44 22–64–24 6.55	M 20 44–22–64 6.25	M 62 20–44–22 5.55	M 18 62–20–44 5.25	M 16 18–62 4.25	M 14 60–16 3.25	M 12 14–60 2.25	M 10 58–12 1.25	M 8 10–58 12.25	M 6 56–8 11.25	M 4 6–56 10.25	M 42 2–54–4 9.25	M 42 2–54–4 8.55	M 42 2–54 8.25	M 52 42–2 7.55	M 52 7.25	M 40 52 6.55	M 52 40 6.25	M 40 5.50		5.30	OW PU
				For time of Main Line Trains between Puyallup Jct. and Bay Street see Time Table No. 21																						
11.20	10.23	9.23	8.23	7.36	7.02	6.32	6.02	5.32	4.32	3.32	2.32	1.32	12.32	11.32	10.32	9.32	9.02	8.32	8.02	7.32	7.02	6.32	5.56		7.68	BS
A 11.35 PM	A 10.35 PM	A 9.35 PM	A 8.35 PM	A 7.50 PM	A 7.15 PM	A 6.45 PM	A 6.15 PM	A 5.45 PM	A 4.50 PM	A 3.50 PM	A 2.50 PM	A 1.50 PM	A 12.50 PM	A 11.50 AM	A 10.50 AM	A 9.50 AM	A 9.15 AM	A 8.45 AM	A 8.15 AM	A 7.45 AM	A 7.15 AM	A 6.45 AM	A 6.10 AM		9.94	X St

SPECIAL RULES

Puyallup Branch Trains be governed by Special Rules Time Table No. 21.

No Train

BRANCH

able A AM 18 NS	Miles From Tacoma	Station Numbers	First Class Daily Ex Sun 202	Daily Ex Sun 204	Daily 206	Daily Ex Sun 208	Daily 210	Daily Ex Sun 212	Daily 214	Daily 216	Daily 218	Daily 220	EASTWARD Daily 222	Daily 224	Daily 226	Daily 228	Daily 230	Daily Ex Sun 232	Daily 234	Daily Ex Sun 236	Daily 238	Daily 240	Daily 242	Daily 244	Daily 246	Daily 248
JP rch Sts.	9.94	A9	A 5.30 AM	A 6.30 AM	A 7.00 AM	A 7.30 AM	A 8.00 AM	A 8.30 AM	A 9.00 AM	A10.00 AM	A11.00 AM	A12.00 Noon	A 1.00 PM	A 2.00 PM	A 3.00 PM	A 4.00 PM	A 5.00 PM	A 5.35 PM	A 6.05 PM	A 6.35 PM	A 7.05 PM	A 7.45 PM	A 8.45 PM	A 9.45 PM	A10.45 PM	A12.22 AM
ON	9.16	A8	5.26	6.25	6.55	7.25	7.55	8.25	8.55	9.55	10.55	11.55	12.55	1.55	2.55	3.55	4.55	5.29	5.58	6.29	6.58	7.40	8.40	9.40	10.40	12.17
D	7.51	A7	5.23	6.20	6.50	7.20	7.50	8.20	8.50	9.50	10.50	11.50	12.50	1.50	2.50	3.50	4.50	5.24	5.54	6.24	6.54	7.35	8.35	9.35	10.35	12.12
	6.38	A6	5.20	M 203 6.18	M 205 6.48	M 207 7.18	M 209 7.48	M 211 8.18	M 213 8.48	9.48	10.48	11.48	12.48	1.48	2.48	3.48	4.48	M 231 5.20	M 233 5.50	M 235 6.20	M 237 6.50	7.33	8.33	9.33	10.33	12.10
R	5.24	A5	5.17	6.15	6.45	7.15	7.45	8.15	8.45	9.45	10.45	11.45	12.45	1.45	2.45	3.45	4.45	5.17	5.47	6.17	6.47	7.30	8.30	9.30	10.30	12.07
JCT.	4.64	5	M 41 51 5.15	M 51 6.13	M 51 1-43 6.43	M 1 43 7.13	M 1 43-53 7.43	M 1 43-53-3 8.13	M 53 3 8.43	M 3 5-55 9.43	M 5 55-7 10.43	M 7 9-57 11.43	M 9 57-11 12.43	M 11 13-59 1.43	M 13 59-15 2.43	M 15 17-61 3.43	M 17 19-45 61 4.43	M 61 19-45 21 5.15	M 19 45-21 63 5.45	M 21 63-23 6.15	M 21 63-23 6.45	M 239 63-23 7.28	M 65 25 8.28	M 65 25-67 9.28	M 67 27 10.28	M 27 69 12.05
						For time of Main Line Trains between Bay Street and Puyallup Jct see Time Table No. 21																				
.	2.26	2A	5.10	6.07	6.37	7.07	7.37	8.07	8.37	9.37	10.37	11.37	12.37	1.37	2.37	3.37	4.37	5.06	5.38	6.08	6.38	7.23	8.23	9.23	10.23	11.57
A Sts.	0	0	L 5.00 AM	M 201 41 L 5.55 AM	M 201 203-41 L 6.25 AM	M 201 203-205 41-51 L 6.55 AM	M 201 203-205 207-41 51 L 7.25 AM	M 201 203-205 207-209 51 L 7.55 AM	M 203 205-207 209-211 51 1-43 L 8.25 AM	M 207 209-211 213-215 1-43-53 L 9.25 AM	M 211 213-215 217-53 3 L10.25 AM	M 215 217-219 3-5-55 L11.25 AM	M 217 219-221 5-55-7 L12.25 PM	M 219 221-223 7-9-57 L 1.25 PM	M 221 223-225 9-57-11 L 2.25 PM	M 223 225-227 11-13-59 L 3.25 PM	M 225 227-229 13-59-15 L 4.25 PM	M 227 229-13 59-15-17 L 4.50 PM	M 227 229-231 15-17-61 L 5.25 PM	M 229 231-233 15-17-61 19-45 L 5.55 PM	M 229 231-233 235-17 19 61-45 L 6.25 PM	M 231 233-235 237-61 19-45-21 L 7.10 PM	M 233 235-237 239-241 21-63-23 L 8.10 PM	M 237 239-241 243-63 23 L 9.10 PM	M 241 243-245 65-25 L10.10 PM	M 245 247 25-67 L11.45 PM

Right By Direction

J. A. MORCKELL,
Chief Dispatcher.

A. L. WELLMAN,
S. NEAL,
L. W. TOWNE,
Dispatchers,

WYE..............Puyallup.
Bay Street.
Spurs..............Firwood.
McAlcer.
Bosworth, 3 Cars.
Sidings............Berryton.
Ardena.
Goldau. 5 Cars

The Puyallup Short Line

PSE 551 crosses the Puyallup River en route to Puyallup. As noted the city was also served by the TR&P which came into town the long way around via Summit and Midland. Although other types of cars were used, the double-ended former TR&P trolleys were the most common. (R.A. Hansen collection) No. 555 is turning from Stewart Street into Meridian. The car will stop just short of the Northern Pacific tracks crossing the street. A portion of an NP locomotive is seen just in front of the trolley. (Author's collection) A sign advertising the Short Line was on this building in Puyallup alongside the NP tracks for several years. (Tacoma Public Library) At right above, looking east at Willow Junction around 1915. The Pacific Highway (Hwy.99) was built between the fence and barn on the left.Compare this photo with another appearing elsewhere in this volume taken in 1927. The small building in the background above the agent is in both photos. Today, it would be south of the Pacific Highway on Valley Avenue (54th Ave.E). Today all traces of the Short Line are gone except for a row of trees to the south. (John C. White) Just below, a pastoral view of Ardena Station on the Short Line. (Author's collection) At right several views of the Tacoma Transit Company vehicles that took over so much of the Short Line's business, the company decided to call in quits in 1922. All appear to be White stages with Goodyear hard rubber tires. (Washington State Historical Society and Tacoma Public Library)

215
PUYALLUP
107
101
TACOMA
PUYALLUP
SUMNER
CIGARS - TOBACCO
TACOMA TRANSIT
TACOMA-PUYALLUP

Above, car No.527 is southbound on Pacific Avenue in Tacoma about 1912. (Lawton Gowey collection) Below, Train No.126 is stopped at Renton Junction in the same year. Motorman Howard Wellman and conductor Horace Rumery pose for the camera. The sign on the Nelsen barn is an advertisement for the Standard Furniture Company which had a store at 2nd and Pine in downtown Seattle. The barn, although in a dilapidated condition, is still at the same location today. If commuter rail is ever a reality this area may become a park & ride lot. (Author's collection)

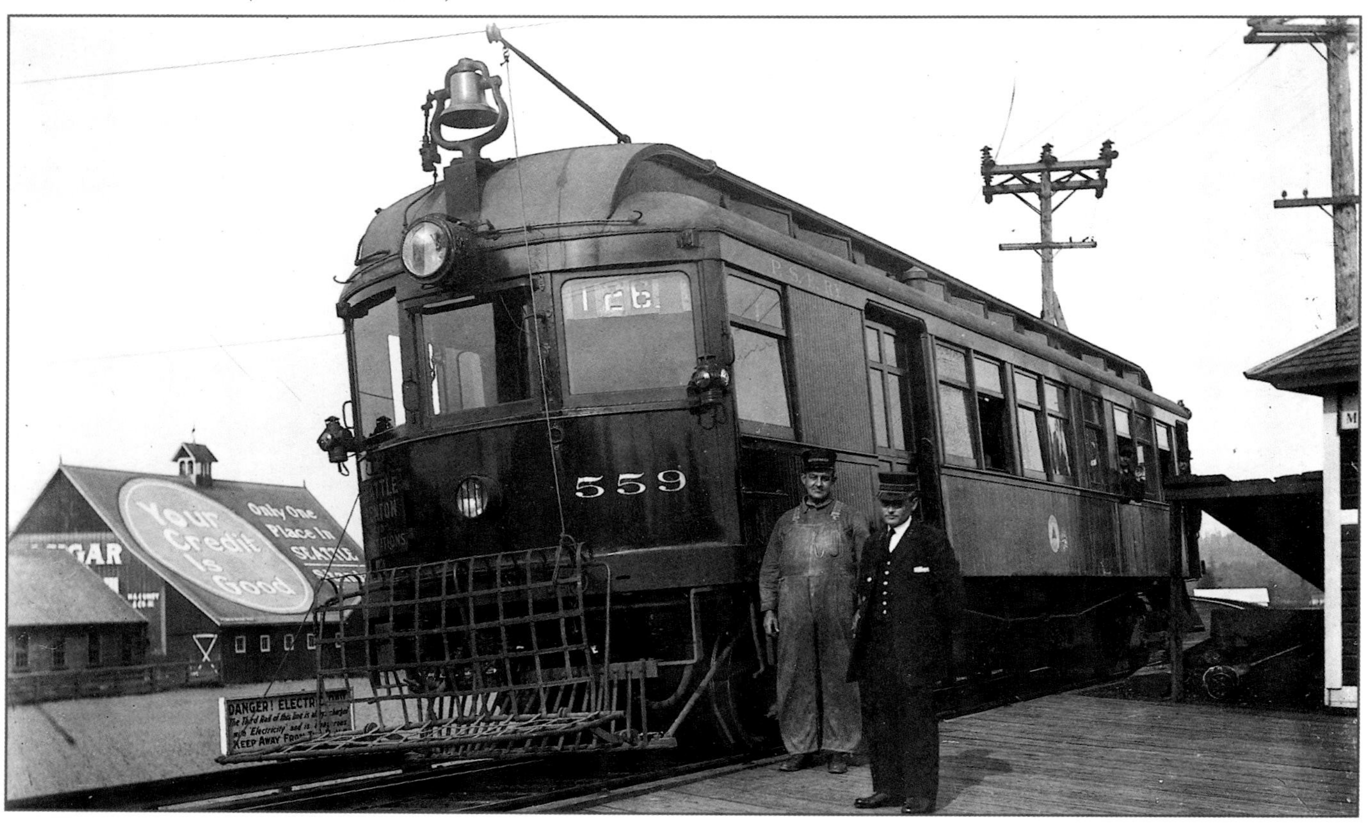

Above, this photo by conductor John C. White, shows southbound motor car No.510 and trailer stopped at Renton Junction in 1912. White was a long time employee and carried a camera from time to time. On the left is new construction for the Valley Highway that will put the interurban out of business 16 years later. Below, Car 561 and trailer are sitting at the end of the Renton branch, also in 1912. (John C. White)

Car No. 559 at the Massachusetts Street carbarn in Seattle about 1915. (Lawton Gowey collection)

An early day Tacoma-Seattle Auto Club tour over the newly planked Pacific Highway at Bluffs. This scene dates from 1913 and you can just imagine what sort of a ride it was with hard rubber tires over wood planks! The Interurban line and Bluffs passenger shelter are in the background at the curve of the highway. The interurban and highway crossed at grade here before it was separated by a steel girder bridge a few years later. The author fondly remembers the curve at Bluffs while on a trip on the interurban in 1928. (Museum of History and Industry)

Looking south on Duwamish Avenue (Airport Way) near Juneau Street after a moderate snowfall in the mid-teens. (Author's collection)

Interurban No.525 with a Renton car behind it is loading at Seattle depot in 1915. (Harold Hill collection)

5¢ CIGAR
EXPERTS
OWL
22
512

Combine No.512 was one of the most photographed cars on the line. Built by the St Louis Car Company in 1907 this beautiful arch-windowed interurban lasted until the end. Above, in a classic pose 512 leads ex-parlor car 529 past the Sears mail order plant at 1st and Lander Street in Seattle. Originally an extra fare car, 529 was motorized and rebuilt into a straight passenger car. (Lawton Gowey collection) Opposite Above, 512 on the head end of train No.22 crossing the Green River at Renton Junction. This photo gives a fairly good view of the working mechanism of the wooden swing bridge. Opposite, 512 is shown again climbing the grade out of the White River Valley high on the hillside in Jovita Canyon. The car is just up the canyon near the entrance of the famous 180 foot tunnel. A larger version of this photograph hung in the Tacoma station for years. After the Interurban quit it was saved by brakeman Al Rockwell and given to the author many years later (Both: Author's collection)

Above, Limited train No.12 in the form of car 529 has just arrived in Seattle at 2:40PM on an overcast day in the mid-1920's (Harold Hill collection) At left, car 525 after having a "meet" with a northbound train, is about to reenter the main line at Milton. A few years later this car suffered an electrical short circuit while passing Jovita. The result was a fire that could not be extinguished. It was pushed onto a side track where it burned to its trucks. (Lawton Gowey collection)

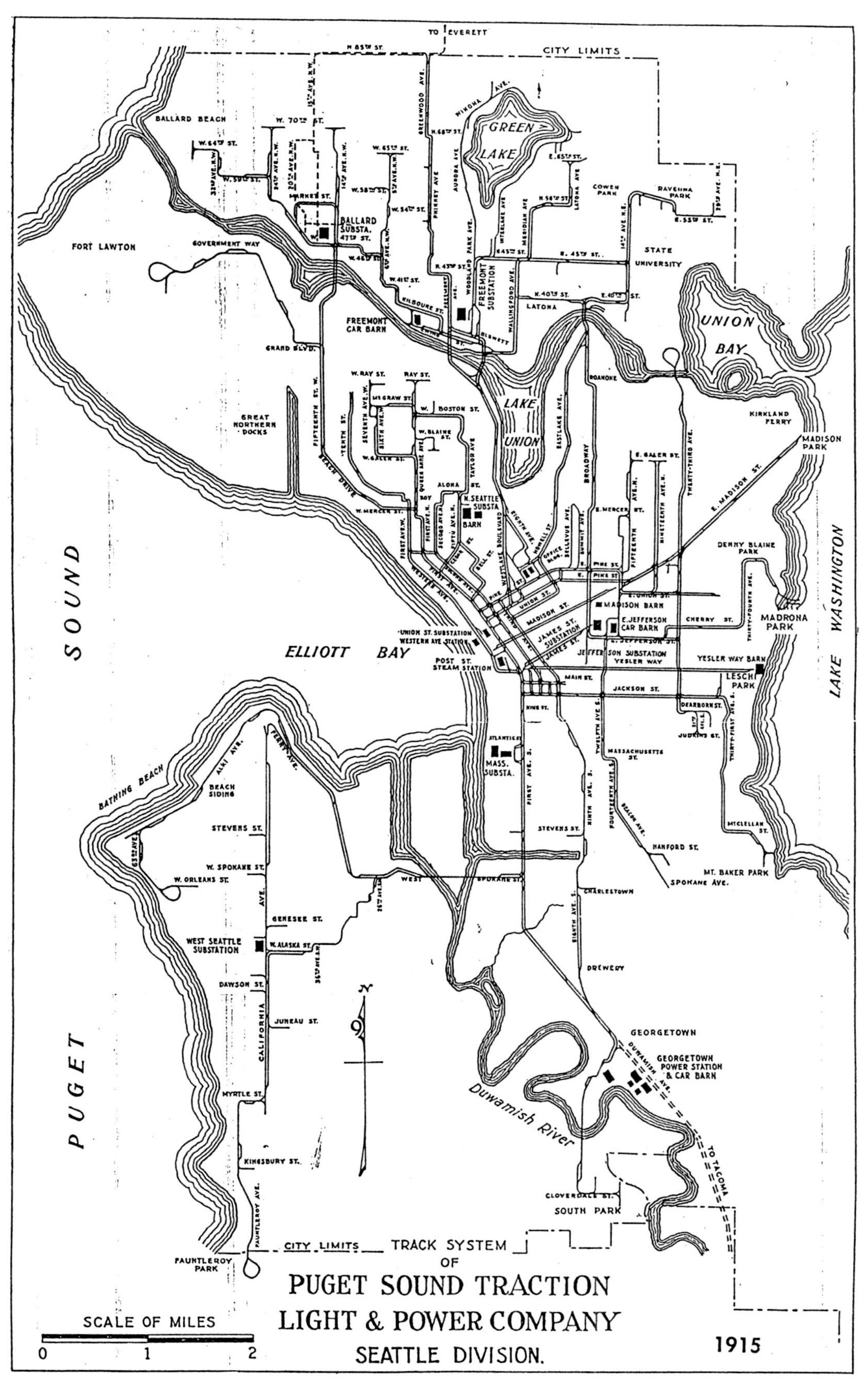
TO EVERETT
CITY LIMITS
GREEN LAKE
BALLARD BEACH
FORT LAWTON
GOVERNMENT WAY
BALLARD SUBSTA.
FREEMONT CAR BARN
FREEMONT SUBSTATION
LATONA
STATE UNIVERSITY
UNION BAY
RAVENNA PARK
COWEN PARK
GRAND BLVD.
GREAT NORTHERN DOCKS
LAKE UNION
KIRKLAND FERRY
MADISON PARK
N. SEATTLE SUBSTA
BARN
DENNY BLAINE PARK
MADISON BARN
E. JEFFERSON CAR BARN
MADRONA PARK
LAKE WASHINGTON
SOUND
ELLIOTT BAY
UNION ST. SUBSTATION
WESTERN AVE. STATION
POST ST. STEAM STATION
JAMES ST. SUBSTATION
JEFFERSON SUBSTATION
YESLER WAY BARN
LESCHI PARK
MASS. SUBSTA.
BATHING BEACH
BEACH SIDING
STEVENS ST.
W. SPOKANE ST.
W. ORLEANS ST.
GENESEE ST.
WEST SEATTLE SUBSTATION
W. ALASKA ST.
DAWSON ST.
JUNEAU ST.
CALIFORNIA AVE.
MYRTLE ST.
KINGSBURY ST.
HANFORD ST.
MT. BAKER PARK
SPOKANE AVE.
CHARLESTOWN
DREWERY
GEORGETOWN
GEORGETOWN POWER STATION & CAR BARN
Duwamish River
TO TACOMA
CLOVERDALE ST.
SOUTH PARK
FAUNTLEROY PARK
PUGET
CITY LIMITS
TRACK SYSTEM
OF
PUGET SOUND TRACTION
LIGHT & POWER COMPANY
SEATTLE DIVISION.
1915
SCALE OF MILES
0
1
2

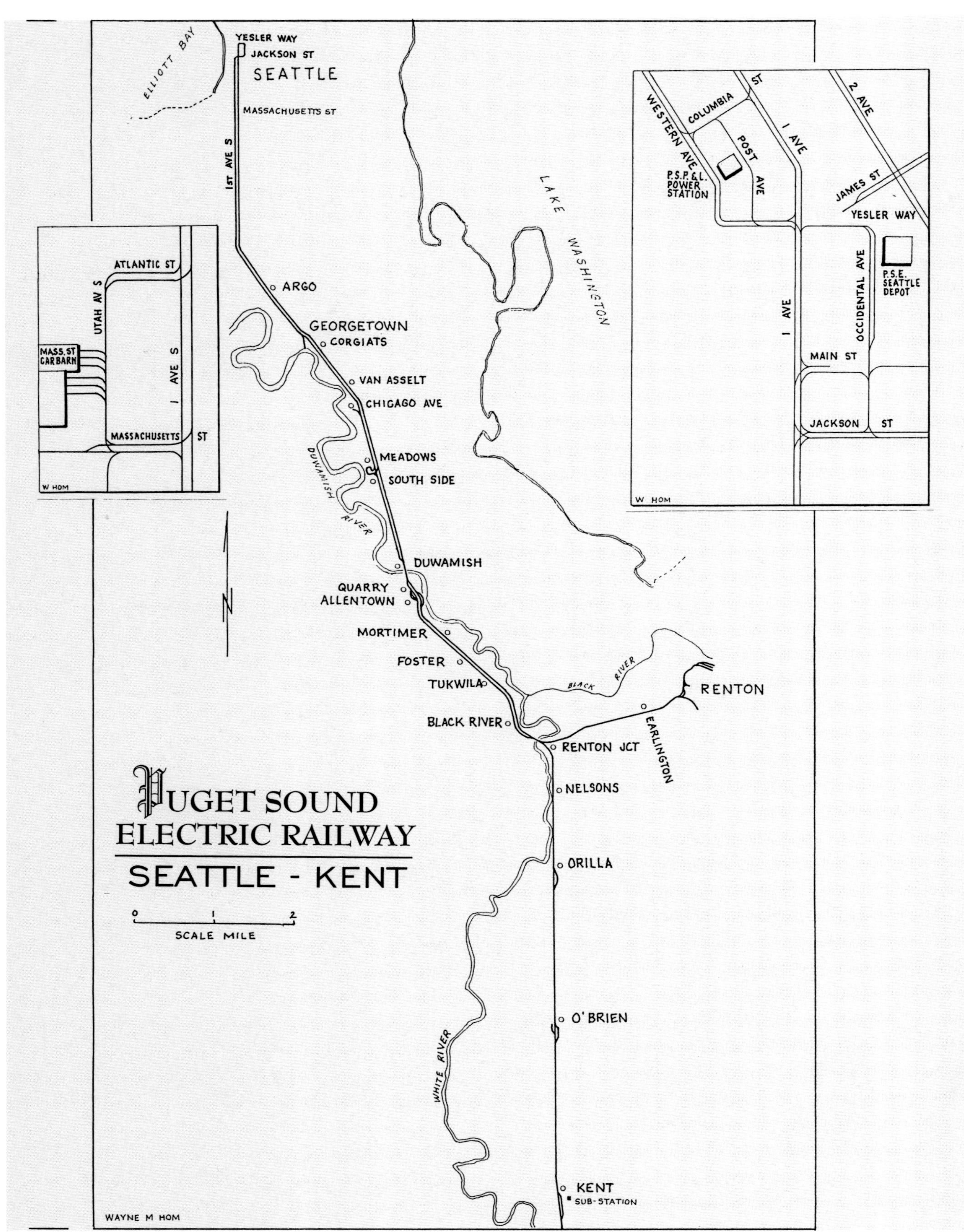

PUGET SOUND
ELECTRIC RAILWAY
SEATTLE - KENT
0 1 2
SCALE MILE
YESLER WAY
JACKSON ST
SEATTLE
ELLIOTT BAY
MASSACHUSETTS ST
1ST AVE S
ARGO
GEORGETOWN
CORGIATS
VAN ASSELT
CHICAGO AVE
MEADOWS
SOUTH SIDE
DUWAMISH RIVER
DUWAMISH
QUARRY
ALLENTOWN
MORTIMER
FOSTER
TUKWILA
BLACK RIVER
RENTON
EARLINGTON
BLACK RIVER
RENTON JCT
NELSONS
ORILLA
O'BRIEN
WHITE RIVER
KENT
SUB-STATION
LAKE WASHINGTON
N
WAYNE M HOM
ATLANTIC ST
UTAH AV S
MASS. ST CAR BARN
I AVE S
MASSACHUSETTS ST
W HOM
WESTERN AVE
COLUMBIA
ST
POST AVE
I AVE
2 AVE
P.S.P.&L. POWER STATION
JAMES ST
YESLER WAY
OCCIDENTAL AVE
P.S.E. SEATTLE DEPOT
MAIN ST
JACKSON ST
W HOM

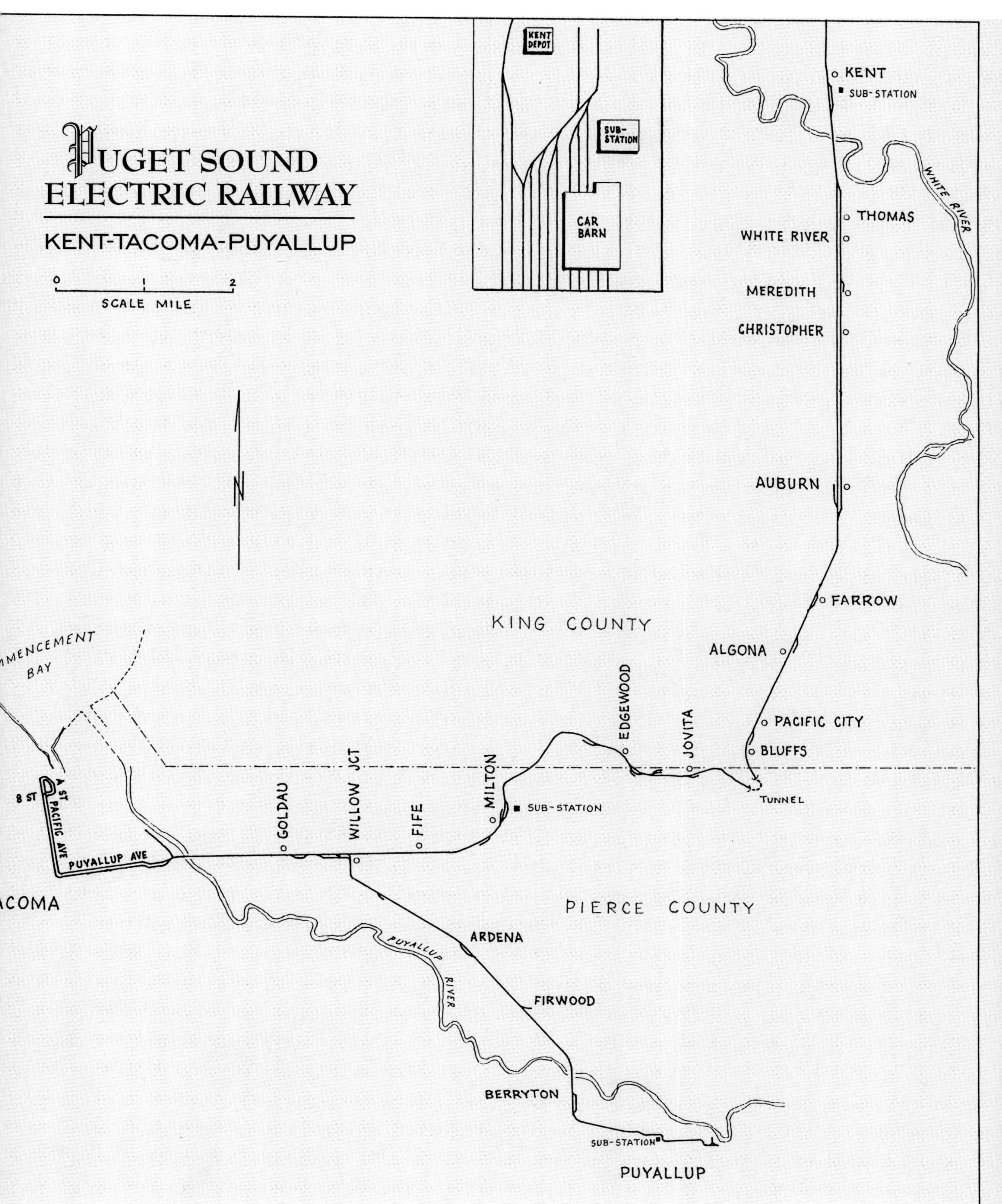
PUGET SOUND
ELECTRIC RAILWAY
KENT-TACOMA-PUYALLUP
0
1
2
SCALE MILE
KENT
DEPOT
SUB-
STATION
CAR
BARN
N
KENT
SUB-STATION
WHITE RIVER
THOMAS
WHITE RIVER
MEREDITH
CHRISTOPHER
AUBURN
FARROW
KING COUNTY
ALGONA
EDGEWOOD
JOVITA
PACIFIC CITY
BLUFFS
TUNNEL
MMENCEMENT
BAY
8 ST
A ST
PACIFIC AVE
PUYALLUP AVE
GOLDAU
WILLOW JCT
FIFE
MILTON
SUB-STATION
ACOMA
PIERCE COUNTY
ARDENA
PUYALLUP
RIVER
FIRWOOD
BERRYTON
SUB-STATION
PUYALLUP

Cars No.518 and 527 at the Seattle depot in the early 1920's. 518 was originally center entrance car No.561 from the Tacoma suburban lines. It was built in 1909 and rebuilt and renumbered a few years later. It was sold in 1929 and became a crew car on a logging road. (Lawton Gowey collection)

At left, a poor but rare view of the station at Auburn taken around 1915. Foster depot (below) was almost at the same location as today's Metro passenger shelter, both located just above the Duwamish River. (Both: Author's collection) Lastly, a charming photo of Frank and Nellie Cummings with their grandmother, Addie Bigelow at Quarry station in 1916. The children's father was a conductor on the line. From the platform of his train, he would signal his wife when he was returning home by sailing a scrap of paper with the time on it into their backyard, which was right next to the tracks. (Courtesy of Frank Cummings)

A great photograph of the Interurban in its heyday. Motorcar No.514 leads a three-car train, bound for Seattle on Tacoma's Pacific Avenue. Just behind the train is a single truck Birney car. Hundreds of these little trolleys were built in the 1920's as a way of saving costs. Yet all they did was delay the inevitable as more and more automobiles appeared on city streets. (Boland photo, courtesy Tacoma Public Library)

3. The Competition

In his book "The Interurban Era" William Middleton called the interurban an "American transportation phenomenon.". It flourished in the early part of the 20th century only to die off, a mere three decades later, a victim of changing tastes and the internal combustion engine. Yet as author Middleton also points out, while the interurban was considered a failure, it helped bridge the gap between the horse and buggy and the modern era of automobiles and super highways.

The interurban's arrival was greeted with optimism and fanfare. This was followed by a few profitable years and then the long slide to the end. Early on, competition from steam railroads and the "jitney" bus cut into the interurbans revenue. In the case of the Seattle-Tacoma Interurban it began with one hand tied behind its back since most of its route was paralleled by three major steam railroads; Northern Pacific, Union Pacific and the Milwaukee Road(C.M.StP&P). The shortline Pacific Coast Railway also followed the Interurban as far as Black River.

Also, the Interurban was faced with competition from other electric railroads, real or otherwise. It seems every other week the local newspapers were reporting some scheme for a new interurban railway. In December 1903, a little more than a year after S-T.I. was completed, Stone & Webster announced plans to build a new line to Tacoma via the shoreline and bluffs overlooking Puget Sound. A short time later, Charles Baker, of Snoqualmie Falls Power Co. and a bitter rival of Stone & Webster, announced plans to compete with S&W by constructing a line between Olympia and Seattle. In September 1905, Fred J. Kerr and associates announced intentions of constructing still another interurban from Seattle to Tacoma. This new line would run via Sunnydale and Des Moines, along Stone's Landing and Jules Gulch in Buena, then on through the Puyallup Indian Reservation into Lincoln Blvd. and on to downtown Tacoma.

Five years later a new company was formed to build an electric line from Kent into Renton to connect with the Seattle, Renton & Southern. The Valley Railroad and Power Company expected construction to begin sometime that summer and eventually build south to Auburn and Tacoma.

To counter, Puget Sound Electric proposed an ambitious plan to spend over $356,000 to upgrade their main line to Tacoma. It called for new stations at Puyallup on the Short Line, Black River, Renton and a new tool house at Bay Street yards in Tacoma. Plans also included a new parlor car, motor car, ten new boxcars and new carbarns at Tacoma and Massachusetts Street in Seattle, plus a second track to Black River.

Later in 1910, backers of a new Mount Hood Railway and Power Company introduced a grand plan they claimed would lead to the "greatest interurban electric system in the Pacific Northwest", linking Portland and Seattle. They envisioned using the grade surveyed by Union Pacific that had been abandoned when UP entered into a ninety year agreement with the Northern Pacific for trackage rights.

However, all these ambitious plans and grand proposals never materialized.

The first hint of things to come came in December 1908 when citizens of Puyallup and Sumner joined forces to get a proposed highway routed through their towns rather than up over the hill through Edgewood. Valley citizens were being urged to donate portions of their property for the highway. Early in 1909, Valley City residents had an energetic meeting at Millgan's Hall to secure a "Macadamized"[2] highway from Seattle. At the same meeting they also demanded a passenger and freight depot be built by the Interurban and the new Chicago, Milwaukee & St. Paul Railroad. By April the interurban company began construction of the new depot and a month later the name of the town was changed to Algona, an Indian name meaning "Valley of Beautiful Flowers".

A few years later, in 1913, a new bus line began offering service between 1st and Virginia in Seattle to Renton via the Duwamish Valley. The first ticket was purchased by Paul Hauser of Renton. Hauser, in turn, gave the ticket to the daughter of E. W. Engel of Tukwila, President of the new bus line. The daughter of the Vice President, George E. Jones, christened the new bus, "Tukwila and Duwamish". The first ticket sold at auction for $13. The total sales for the day amounted to $100.

Meanwhile earlier in the year, the state highway between Auburn and the Pierce County line was paved making it more reliable to new auto and bus traffic. This was the final section paved between Auburn and Tacoma.

In a special report to the Washington State Public Service Commission, PSE stated that local business had dropped more than $125,000 between August 1913 and August 1914, due to competition from local buses while through passenger business remained about the same.

Also in August 1913, PSE announced further improvements in the form of a $60,000 block signal system as well as upgrading its tracks.

Still, by 1914, jitney buses were skimming business from the company's local streetcar lines. The jitneys were taking the profitable short haul business and leaving the company with the unprofitable long hauls. One of Seattle's city council members stated he could see the end of electric cars in favor of motor buses. The traction company countered by threatening to suspend service on its longer lines, such as the Alki, Ballard North and Fauntleroy.

Finally in May 1918, the city of Seattle made an offer of $15 million dollars for the city's streetcar system. PSTL&P president A. W. Leonard accepted although he declared the city was getting a bargain since the market value was more like 25 million. Mayor Ole Hanson signed the deal on January 1, 1919

During the war years Leonard was elected president of the Army-Navy Club and appointed chairman of a commission to establish a soldier

Opposite above, Limited train No. 175 is stopped at the Kent depot in 1915 (Lawton Gowey collection) Right, looking north to Algona, year unknown. Originally called Valley City it was renamed Algona a few years after the interurban began. It was one of a number of small communities devastated when the interurban shut down. (Boland photo, Tacoma Public Library)

2. Asphalt paving was known by this name in honor of its inventor, John McAdam

175

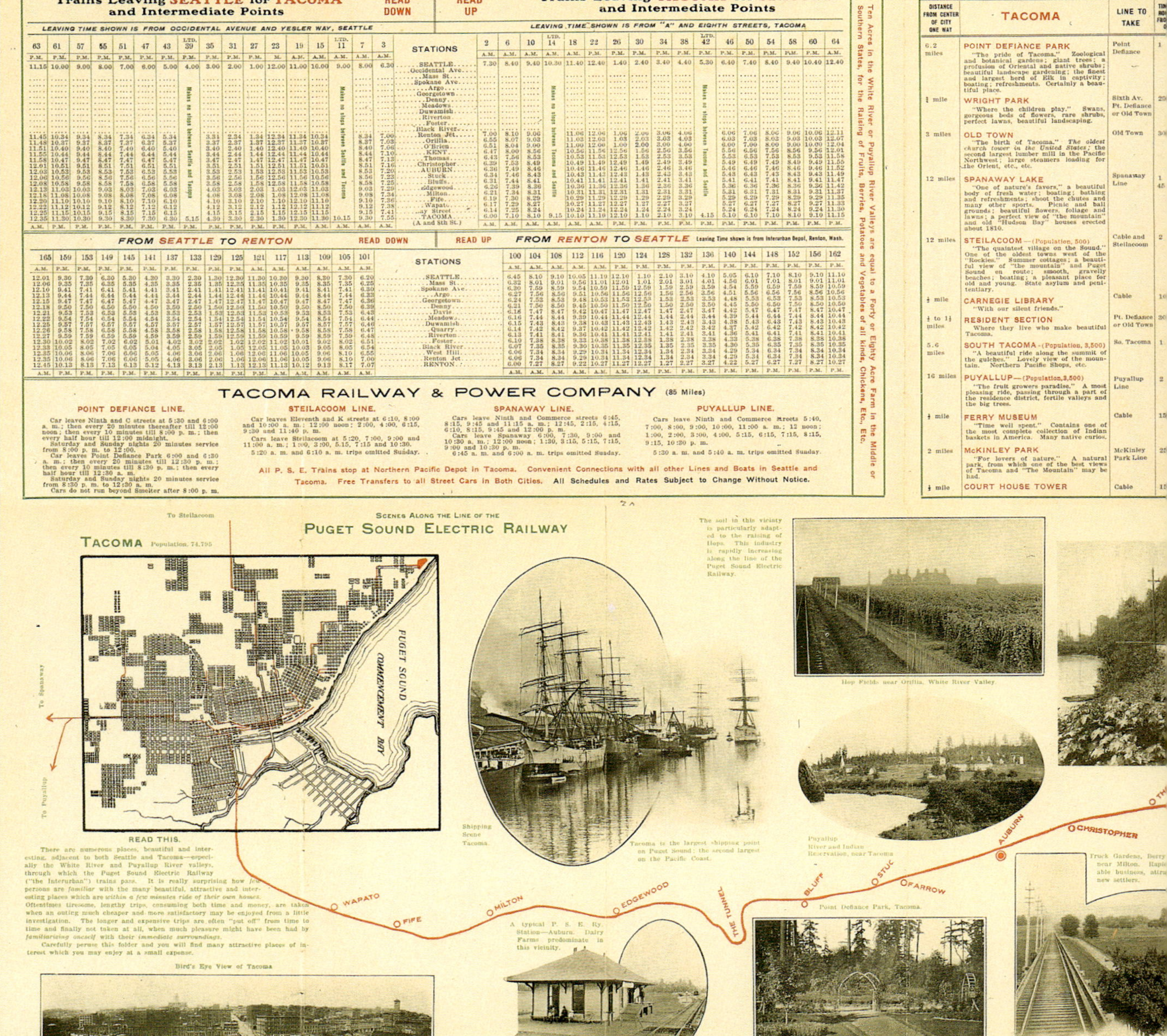

Seattle to Tacoma, 36 miles

Puget Sound Electric Railway. Daily Time Card

Total Mileage 52 Miles

Trains Leaving SEATTLE for TACOMA and Intermediate Points — READ DOWN

LEAVING TIME SHOWN IS FROM OCCIDENTAL AVENUE AND YESLER WAY, SEATTLE

Stations	63	61	57	55	51	47	43	LTD. 39	35	31	27	23	19	15	LTD. 11	7	3
	P.M.	P.M.	P.M.	P.M.	P.M.	P.M.	P.M.	P.M.	P.M.	P.M.	P.M.	M.	A.M.	A.M.	A.M.	A.M.	A.M.
SEATTLE	11.15	10.00	9.00	8.00	7.00	6.00	5.00	4.00	3.00	2.00	1.00	12.00	11.00	10.00	9.00	8.00	6.30
Occidental Ave.	.....	.....	.....	.....	.....	.....	.....	Makes no stops between Seattle and Tacoma	.....	.....	.....	.....	.....	.....	Makes no stops between Seattle and Tacoma	.....	.....
Mass St.	.....	.....	.....	.....	.....	.....	.....		.....	.....	.....	.....	.....	.....		.....	.....
Spokane Ave.	.....	.....	.....	.....	.....	.....	.....		.....	.....	.....	.....	.....	.....		.....	.....
Argo	.....	.....	.....	.....	.....	.....	.....		.....	.....	.....	.....	.....	.....		.....	.....
Georgetown	.....	.....	.....	.....	.....	.....	.....		.....	.....	.....	.....	.....	.....		.....	.....
Denny	.....	.....	.....	.....	.....	.....	.....		.....	.....	.....	.....	.....	.....		.....	.....
Meadows	.....	.....	.....	.....	.....	.....	.....		.....	.....	.....	.....	.....	.....		.....	.....
Duwamish	.....	.....	.....	.....	.....	.....	.....		.....	.....	.....	.....	.....	.....		.....	.....
Riverton	.....	.....	.....	.....	.....	.....	.....		.....	.....	.....	.....	.....	.....		.....	.....
Foster	.....	.....	.....	.....	.....	.....	.....		.....	.....	.....	.....	.....	.....		.....	.....
Black River	.....	.....	.....	.....	.....	.....	.....		.....	.....	.....	.....	.....	.....		.....	.....
Renton Jct.	11.45	10.34	9.34	8.34	7.34	6.34	5.34		3.34	2.34	1.34	12.34	11.34	10.34		8.34	7.00
Orillia	11.48	10.37	9.37	8.37	7.37	6.37	5.37		3.37	2.37	1.37	12.37	11.37	10.37		8.37	7.03
O'Brien	11.51	10.40	9.40	8.40	7.40	6.40	5.40		3.40	2.40	1.40	12.40	11.40	10.40		8.40	7.06
KENT	11.55	10.44	9.44	8.44	7.44	6.44	5.44		3.44	2.44	1.44	12.44	11.44	10.44		8.44	7.10
Thomas	11.58	10.47	9.47	8.47	7.47	6.47	5.47		3.47	2.47	1.47	12.47	11.47	10.47		8.47	7.13
Christopher	12.01	10.51	9.51	8.51	7.51	6.51	5.51		3.51	2.51	1.51	12.51	11.51	10.51		8.51	7.16
AUBURN	12.03	10.53	9.53	8.53	7.53	6.53	5.53		3.53	2.53	1.53	12.53	11.53	10.53		8.53	7.20
Stuck	12.06	10.56	9.56	8.56	7.56	6.56	5.56		3.56	2.56	1.56	12.56	11.56	10.56		8.56	7.22
Bluffs	12.08	10.58	9.58	8.58	7.58	6.58	5.58		3.58	2.58	1.58	12.58	11.58	10.58		8.58	7.25
Edgewood	12.13	11.03	10.03	9.03	8.03	7.03	6.03		4.03	3.03	2.03	1.03	12.03	11.03		9.03	7.29
Milton	12.18	11.08	10.08	9.08	8.08	7.08	6.08		4.08	3.08	2.08	1.08	12.08	11.08		9.08	7.34
Fife	12.20	11.10	10.10	9.10	8.10	7.10	6.10		4.10	3.10	2.10	1.10	12.10	11.10		9.10	7.36
Wapato	12.22	11.12	10.12	9.12	8.12	7.12	6.12		4.12	3.12	2.12	1.12	12.12	11.12		9.12	7.38
...ay Street	12.25	11.15	10.15	9.15	8.15	7.15	6.15		4.15	3.15	2.15	1.15	12.15	11.15		9.15	7.41
TACOMA (A and 8th St.)	12.35	11.30	10.30	9.30	8.30	7.30	6.30	5.15	4.30	3.30	2.30	1.30	12.30	11.30	10.15	9.30	7.55
	A.M.	P.M.	P.M.	P.M.	P.M.	P.M.	P.M.	P.M.	P.M.	P.M.	P.M.	P.M.	P.M.	A.M.	A.M.	A.M.	A.M.

Trains Leaving TACOMA for SEATTLE and Intermediate Points — READ UP

LEAVING TIME SHOWN IS FROM "A" AND EIGHTH STREETS, TACOMA

Stations	2	6	10	LTD. 14	18	22	26	30	34	38	LTD. 42	46	50	54	58	60	64
	A.M.	A.M.	A.M.	A.M.	A.M.	P.M.	P.M.	P.M.	P.M.	P.M.	P.M.	P.M.	P.M.	P.M.	P.M.	P.M.	A.M.
SEATTLE	7.30	8.40	9.40	10.30	11.40	12.40	1.40	2.40	3.40	4.40	5.30	6.40	7.40	8.40	9.40	10.40	12.40
Occidental Ave.	.....	.....	.....	Makes no stops between Tacoma and Seattle	.....	.....	.....	.....	.....	.....	Makes no stops between Tacoma and Seattle	.....	.....	.....	.....	.....	.....
Mass St.	.....	.....	.....		.....	.....	.....	.....	.....	.....		.....	.....	.....	.....	.....	.....
Spokane Ave.	.....	.....	.....		.....	.....	.....	.....	.....	.....		.....	.....	.....	.....	.....	.....
Argo	.....	.....	.....		.....	.....	.....	.....	.....	.....		.....	.....	.....	.....	.....	.....
Georgetown	.....	.....	.....		.....	.....	.....	.....	.....	.....		.....	.....	.....	.....	.....	.....
Denny	.....	.....	.....		.....	.....	.....	.....	.....	.....		.....	.....	.....	.....	.....	.....
Meadows	.....	.....	.....		.....	.....	.....	.....	.....	.....		.....	.....	.....	.....	.....	.....
Duwamish	.....	.....	.....		.....	.....	.....	.....	.....	.....		.....	.....	.....	.....	.....	.....
Riverton	.....	.....	.....		.....	.....	.....	.....	.....	.....		.....	.....	.....	.....	.....	.....
Foster	.....	.....	.....		.....	.....	.....	.....	.....	.....		.....	.....	.....	.....	.....	.....
Black River	.....	.....	.....		.....	.....	.....	.....	.....	.....		.....	.....	.....	.....	.....	.....
Renton Jct.	7.00	8.10	9.06		11.06	12.06	1.06	2.06	3.06	4.06		6.06	7.06	8.06	9.06	10.06	12.11
Orillia	6.55	8.07	9.03		11.03	12.03	1.03	2.03	3.03	4.03		6.03	7.03	8.03	9.03	10.03	12.07
O'Brien	6.51	8.04	9.00		11.00	12.00	1.00	2.00	3.00	4.00		6.00	7.00	8.00	9.00	10.00	12.04
KENT	6.47	8.00	8.56		10.56	11.56	12.56	1.56	2.56	3.56		5.56	6.56	7.56	8.56	9.56	12.01
Thomas	6.43	7.56	8.53		10.53	11.53	12.53	1.53	2.53	3.53		5.53	6.53	7.53	8.53	9.53	11.58
Christopher	6.39	7.53	8.49		10.49	11.49	12.49	1.49	2.49	3.49		5.49	6.49	7.49	8.49	9.49	11.55
AUBURN	6.36	7.49	8.46		10.46	11.46	12.46	1.46	2.46	3.46		5.46	6.46	7.46	8.46	9.46	11.52
Stuck	6.34	7.46	8.43		10.43	11.43	12.43	1.43	2.43	3.43		5.43	6.43	7.43	8.43	9.43	11.49
Bluffs	6.31	7.44	8.41		10.41	11.41	12.41	1.41	2.41	3.41		5.41	6.41	7.41	8.41	9.41	11.47
Edgewood	6.26	7.39	8.36		10.36	11.36	12.36	1.36	2.36	3.36		5.36	6.36	7.36	8.36	9.36	11.42
Milton	6.21	7.34	8.31		10.31	11.31	12.31	1.31	2.31	3.31		5.31	6.31	7.31	8.31	9.31	11.37
Fife	6.19	7.30	8.29		10.29	11.29	12.29	1.29	2.29	3.29		5.29	6.29	7.29	8.29	9.29	11.35
Wapato	6.17	7.29	8.27		10.27	11.27	12.27	1.27	2.27	3.27		5.27	6.27	7.27	8.27	9.27	11.33
...ay Street	6.14	7.25	8.24		10.24	11.24	12.24	1.24	2.24	3.24		5.24	6.24	7.24	8.24	9.24	11.30
TACOMA (A and 8th St.)	6.00	7.10	8.10	9.15	10.10	11.10	12.10	1.10	2.10	3.10	4.15	5.10	6.10	7.10	8.10	9.10	11.15
	A.M.	A.M.	A.M.	A.M.	A.M.	A.M.	P.M.	P.M.	P.M.	P.M.	P.M.	P.M.	P.M.	P.M.	P.M.	P.M.	P.M.

FROM SEATTLE TO RENTON — READ DOWN

Stations	165	159	153	149	145	141	137	133	129	125	121	117	113	109	105	101
	A.M.	P.M.	P.M.	P.M.	P.M.	P.M.	P.M.	P.M.	P.M.	P.M.	A.M.	A.M.	A.M.	A.M.	A.M.	A.M.
SEATTLE	12.01	9.30	7.30	6.30	5.30	4.30	3.30	2.30	1.30	12.30	11.30	10.30	9.30	8.30	7.30	6.20
Mass St.	12.06	9.35	7.35	6.35	5.35	4.35	3.35	2.35	1.35	12.35	11.35	10.35	9.35	8.35	7.35	6.25
Spokane Ave.	12.10	9.41	7.41	6.41	5.41	4.41	3.41	2.41	1.41	12.41	11.41	10.41	9.41	8.41	7.41	6.30
Argo	12.13	9.44	7.44	6.44	5.44	4.44	3.44	2.44	1.44	12.44	11.44	10.44	9.44	8.44	7.44	6.33
Georgetown	12.15	9.47	7.47	6.47	5.47	4.47	3.47	2.47	1.47	12.47	11.47	10.47	9.47	8.47	7.47	6.36
Denny	12.18	9.50	7.50	6.50	5.50	4.50	3.50	2.50	1.50	12.50	11.50	10.50	9.50	8.50	7.50	6.39
Davis	12.20	9.52	7.52	6.52	5.52	4.52	3.52	2.52	1.52	12.52	11.52	10.52	9.52	8.52	7.52	6.42
Meadow	12.22	9.54	7.54	6.54	5.54	4.54	3.54	2.54	1.54	12.54	11.54	10.54	9.54	8.54	7.54	6.44
Duwamish	12.25	9.57	7.57	6.57	5.57	4.57	3.57	2.57	1.57	12.57	11.57	10.57	9.57	8.57	7.57	6.46
Quarry	12.26	9.58	7.58	6.58	5.58	4.58	3.58	2.58	1.58	12.58	11.58	10.58	9.58	8.58	7.58	6.47
Riverton	12.27	9.59	7.59	6.59	5.59	4.59	3.59	2.59	1.59	12.59	11.59	10.59	9.59	8.59	7.59	6.48
Foster	12.30	10.02	8.02	7.02	6.02	5.02	4.02	3.02	2.02	1.02	12.02	11.02	10.02	9.02	8.02	6.51
Black River	12.33	10.05	8.05	7.05	6.05	5.05	4.05	3.05	2.05	1.05	12.05	11.05	10.05	9.05	8.05	6.54
West Hill	12.35	10.06	8.06	7.06	6.06	5.06	4.06	3.06	2.06	1.06	12.06	11.06	10.06	9.06	8.06	6.55
Renton Jct.	12.40	10.10	8.10	7.10	6.10	5.10	4.10	3.10	2.10	1.10	12.10	11.10	10.10	9.10	8.10	7.00
RENTON	12.45	10.13	8.13	7.13	6.13	5.12	4.13	3.13	2.13	1.13	12.13	11.13	10.12	9.13	8.17	7.07
	A.M.	P.M.	P.M.	P.M.	P.M.	P.M.	P.M.	P.M.	P.M.	P.M.	P.M.	A.M.	A.M.	A.M.	A.M.	A.M.

FROM RENTON TO SEATTLE — READ UP

Leaving Time shown is from Interurban Depot, Renton, Wash.

Stations	100	104	108	112	116	120	124	128	132	136	140	144	148	152	156	162
	A.M.	A.M.	A.M.	A.M.	A.M.	P.M.	P.M.	P.M.	P.M.	P.M.	P.M.	P.M.	P.M.	P.M.	P.M.	P.M.
SEATTLE	6.45	8.10	9.10	10.05	11.10	12.10	1.10	2.10	3.10	4.10	5.05	6.10	7.10	8.10	9.10	11.10
Mass St.	6.32	8.01	9.01	9.56	11.01	12.01	1.01	2.01	3.01	4.01	4.56	6.01	7.01	8.01	9.01	11.01
Spokane Ave.	6.30	7.59	8.59	9.54	10.59	11.59	12.59	1.59	2.59	3.59	4.54	5.59	6.59	7.59	8.59	10.59
Argo	6.27	7.56	8.56	9.51	10.56	11.56	12.56	1.56	2.56	3.56	4.51	5.56	6.56	7.56	8.56	10.56
Georgetown	6.24	7.53	8.53	9.48	10.53	11.53	12.53	1.53	2.53	3.53	4.48	5.53	6.53	7.53	8.53	10.53
Denny	6.21	7.50	8.50	9.45	10.50	11.50	12.50	1.50	2.50	3.50	4.45	5.50	6.50	7.50	8.50	10.50
Davis	6.18	7.47	8.47	9.42	10.47	11.47	12.47	1.47	2.47	3.47	4.42	5.47	6.47	7.47	8.47	10.47
Meadow	6.16	7.44	8.44	9.39	10.44	11.44	12.44	1.44	2.44	3.44	4.39	5.44	6.44	7.44	8.44	10.44
Duwamish	6.15	7.43	8.43	9.38	10.43	11.43	12.43	1.43	2.43	3.43	4.38	5.43	6.43	7.43	8.43	10.43
Quarry	6.14	7.42	8.42	9.37	10.42	11.42	12.42	1.42	2.42	3.42	4.37	5.42	6.42	7.42	8.42	10.42
Riverton	6.13	7.41	8.41	9.36	10.41	11.41	12.41	1.41	2.41	3.41	4.36	5.41	6.41	7.41	8.41	10.41
Foster	6.10	7.38	8.38	9.33	10.38	11.38	12.38	1.38	2.38	3.38	4.33	5.38	6.38	7.38	8.38	10.38
Black River	6.07	7.35	8.35	9.30	10.35	11.35	12.35	1.35	2.35	3.35	4.30	5.35	6.35	7.35	8.35	10.35
West Hill	6.06	7.34	8.34	9.29	10.34	11.34	12.34	1.34	2.34	3.34	4.29	5.34	6.34	7.34	8.34	10.34
Renton Jct.	6.06	7.34	8.34	9.29	10.34	11.34	12.34	1.34	2.34	3.34	4.29	5.34	6.34	7.34	8.34	10.34
RENTON	6.00	7.27	8.27	9.22	10.27	11.27	12.27	1.27	2.27	3.27	4.22	5.27	6.27	7.27	8.27	10.27
	A.M.	A.M.	A.M.	A.M.	A.M.	A.M.	P.M.	P.M.	P.M.	P.M.	P.M.	P.M.	P.M.	P.M.	P.M.	P.M.

TACOMA RAILWAY & POWER COMPANY (85 Miles)

POINT DEFIANCE LINE.

Car leaves Ninth and C streets at 5:30 and 6:00 a. m.; then every 20 minutes thereafter till 12:00 noon; then every 10 minutes till 8:00 p. m.; then every half hour till 12:00 midnight.
Saturday and Sunday nights 20 minutes service from 8:00 p. m. to 12:00.
Car leaves Point Defiance Park 6:00 and 6:30 a. m.; then every 20 minutes till 12:30 p. m.; then every 10 minutes till 8:30 p. m.; then every half hour till 12:30 a. m.
Saturday and Sunday nights 20 minutes service from 8:30 p. m. to 12:30 a. m.
Cars do not run beyond Smelter after 8:00 p. m.

STEILACOOM LINE.

Car leaves Eleventh and K streets at 6:10, 8:00 and 10:00 a. m.; 12:00 noon; 2:00, 4:00, 6:15, 9:30 and 11:40 p. m.
Cars leave Steilacoom at 5:20, 7:00, 9:00 and 11:00 a. m.; 1:00, 3:00, 5:15, 7:15 and 10:30.
5:20 a. m. and 6:10 a. m. trips omitted Sunday.

SPANAWAY LINE.

Cars leave Ninth and Commerce streets 6:45, 8:15, 9:45 and 11:15 a. m.; 12:45, 2:15, 4:15, 6:10, 8:15, 9:45 and 12:00 p. m.
Cars leave Spanaway 6:00, 7:30, 9:00 and 10:30 a. m.; 12:00 noon; 1:30, 3:15, 5:15, 7:15, 9:00 and 10:30 p. m.
6:45 a. m. and 6:00 a. m. trips omitted Sunday.

PUYALLUP LINE.

Cars leave Ninth and Commerce streets 5:40, 7:00, 8:00, 9:00, 10:00, 11:00 a. m.; 12 noon; 1:00, 2:00, 3:00, 4:00, 5:15, 6:15, 7:15, 8:15, 9:15, 10:30 p. m.
5:30 a. m. and 5:40 a. m. trips omitted Sunday.

All P. S. E. Trains stop at Northern Pacific Depot in Tacoma. Convenient Connections with all other Lines and Boats in Seattle and Tacoma. Free Transfers to all Street Cars in Both Cities. All Schedules and Rates Subject to Change Without Notice.

Five or Ten Acres in the White River or Puyallup River Valleys are equal to a Forty or Eighty Acre Farm in the Middle or Southern States, for the Raising of Fruits, Berries, Potatoes and Vegetables of all kinds, Chickens, Etc., Etc.

"The Scenic Route of Puget Sound"

Places of Special Interest and How to Reach Them

Distance from Center of City One Way	TACOMA	Line to Take
6.2 miles	**POINT DEFIANCE PARK** "The pride of Tacoma." Zoological and botanical gardens; giant trees; a profusion of Oriental and native shrubs; beautiful landscape gardening; the finest and largest herd of Elk in captivity; boating; refreshments. Certainly a beautiful place.	Point Defiance
1 mile	**WRIGHT PARK** "Where the children play." Swans, gorgeous beds of flowers, rare shrubs, perfect lawns, beautiful landscaping.	Sixth Av. Pt. Defiance or Old Town
3 miles	**OLD TOWN** "The birth of Tacoma." *The oldest church tower in the United States*; the second largest lumber mill in the Pacific Northwest; large steamers loading for the Orient, etc., etc.	Old Town
12 miles	**SPANAWAY LAKE** "One of nature's favors," a beautiful body of fresh water; boating; bathing and refreshments; shoot the chutes and many other sports. Picnic and ball grounds; beautiful flowers, foliage and lawns; a perfect view of "the mountain" and old "Hudson Bay" houses erected about 1810.	Spanaway Line
12 miles	**STEILACOOM—(Population, 500)** "The quaintest village on the Sound." One of the oldest towns west of the "Rockies." Summer cottages; a beautiful view of "the mountain" and Puget Sound en route; smooth, gravelly beaches; boating; a pleasant place for old and young. State asylum and penitentiary.	Cable and Steilacoom
½ mile	**CARNEGIE LIBRARY** "With our silent friends."	Cable
½ to 1½ miles	**RESIDENT SECTION** Where they live who make beautiful Tacoma.	Pt. Defiance or Old Town
5.6 miles	**SOUTH TACOMA—(Population, 3,500)** "A beautiful ride along the summit of the gulches." Lovely view of the mountain. Northern Pacific Shops, etc.	So. Tacoma
10 miles	**PUYALLUP—(Population, 3,500)** "The fruit growers paradise." A most pleasing ride, passing through a part of the residence district, fertile valleys and the big trees.	Puyallup Line
½ mile	**FERRY MUSEUM** "Time well spent." Contains one of the most complete collection of Indian baskets in America. Many native curios.	Cable
2 miles	**McKINLEY PARK** "For lovers of nature." A natural park, from which one of the best views of Tacoma and "The Mountain" may be had.	McKinley Park Line
½ mile	**COURT HOUSE TOWER**	Cable

Scenes Along the Line of the

Puget Sound Electric Railway

READ THIS.

There are numerous places, beautiful and interesting, adjacent to both Seattle and Tacoma—especially the White River and Puyallup River valleys, through which the Puget Sound Electric Railway ("the Interurban") trains pass. It is really surprising how *few* persons are *familiar* with the many beautiful, attractive and interesting places which are *within a few minutes ride of their own homes.* Oftentimes tiresome, lengthy trips, consuming both time and money, are taken when an outing much cheaper and more satisfactory may be enjoyed from a little investigation. The longer and expensive trips are often "put off" from time to time and finally not taken at all, when much pleasure might have been had by *familiarizing oneself* with their *immediate surroundings.*

Carefully peruse this folder and you will find many attractive places of interest which you may enjoy at a small expense.

Bird's Eye View of Tacoma

Shipping Scene Tacoma.

Tacoma is the largest shipping point on Puget Sound; the second largest on the Pacific Coast.

The soil in this vicinity is particularly adapted to the raising of Hops. This industry is rapidly increasing along the line of the Puget Sound Electric Railway.

Hop Fields near Orillia, White River Valley.

Puyallup River and Indian Reservation, near Tacoma

A typical P. S. E. Ry. Station—Auburn. Dairy Farms predominate in this vicinity.

Point Defiance Park, Tacoma.

Truck Gardens, Berry ... near Milton. Rapid... able business, attra... new settlers.

"The Scenic Route of Puget Sound"

Places of Special Interest and How to Reach Them

SEATTLE	LINE TO TAKE	TIME CONSUMED ROUND TRIP FROM CENTER OF CITY
Assay Office	James Street	10 mins.
Ballard	Ballard	1 hr.20 mins.
Beacon Hill	Beacon Hill James Street	40 mins.
Court House	James Street	15 mins.
Capital Hill	Capital Hill	1 hr.
Denny Blaine Park	James St. & Madrona	1 hr.
Fort Lawton	Ft. Lawton	1 hr.30 mins.
Green Lake	Green Lake	1 hr.30 mins.
Georgetown	South Seattle	1 hr.
Great Northern Docks	Ballard	40 mins.
Kinnear Park	Kinnear Park	45 mins.
Leschi Park	Yesler Way	35 mins.
View of City from Washington Hotel	Yesler Way	15 mins.
Lake Union	Green Lake	30 mins.
Lake Washington	Fremont, Ballard Madison Street	1 hr.
Lake View Cemetery	Broadway & Pike	1 hr.
Madison Park	Madison Street	1 hr.
Madrona Park	Madrona & James	1 hr.
Moran's Ship Yard	1st Ave. South	30 mins.
Queen Ann Hill	Queen Anne Lines	1 hr.
Ravenna Park	University	1 hr.30 mins.
Race Tracks	Interurban	1 hr.30 mins.
University of Washington	University	1 hr.
Volunteer Park	Broadway & Pike	1 hr.
Woodland Park	Green Lake	1 hr.30 mins.

PUGET SOUND ELECTRIC RAILWAY

FROM SEATTLE ROUND TRIP	PASSENGER RATES	FROM TACOMA ONE WAY	FROM TACOMA ROUND TRIP
	Seattle	$ .60	$1.00
$ .15	Georgetown	.60	1.00
.25	Renton Junction	.65	1.00
.25	Renton	.60	1.00
.45	Orillia	.50	.90
.50	O'Brien	.45	.90
.50	Kent	.35	.65
.60	Thomas	.35	.60
.70	Christopher	.25	.60
.75	Auburn	.25	.45
.95	Bluffs	.25	.45
.95	Edgewood	.15	.25
1.00	Milton	.10	.15
1.00	Fife	.10	.15
1.00	Wapato	.10	.15
1.00	Tacoma		

TACOMA RAILWAY AND POWER COMPANY

FROM TACOMA TO	ONE WAY	ROUND TRIP
Spanaway	$ .20	$ .25
Puyallup	.20	.25
Steilacoom	.20	.25
South Tacoma	.05	.10
Point Defiance	.05	.10

*Round Trip Sundays during Summer Months to Spanaway, 15c.

PUGET SOUND ELECTRIC RAILWAY

Passenger Rates and Connections

AT TACOMA

NOTE: Puget Sound Electric Railway trains leave Seattle every hour, stopping at Northern Pacific Station in Tacoma. Close connections may be made with south-bound trains.

Rates from Seattle via P. S. E. Ry. to Tacoma, and via N. P. Ry. to:

Olympia	$1.55	Ocosta	$3.75
Montesano	3.00	Centralia	2.05
Aberdeen	3.25	Chehalis	2.20
Hoquiam	3.45	South Bend	3.95
Cosmopolis	3.45	Portland	4.95

TACOMA EASTERN RAILROAD.

Tacoma to Longmire's Springs (in Paradise Valley, near Mt. Rainier or Mt. Tacoma), Round Trip, including 13 miles of stage.......$6.00

From Seattle via P. S. E. Ry., $7.00. Close connections—P. S. E. trains stopping at Tacoma Eastern station in Tacoma.

OLYMPIA & TACOMA NAVIGATION CO.

Passengers leaving Seattle at 10:00 a. m. and 2:00 p. m. make close connections in Tacoma with Olympia & Tacoma Navigation Co's. steamer "Greyhound" for Olympia.

Fare Seattle to Olympia, via P. S. E. Ry.................$1.00

AT SEATTLE

GREAT NORTHERN RAILWAY CO.

Rates via Puget Sound Electric Railway and Great Northern Railway:

Tacoma to Anacortes	$3.10	Tacoma to Everett	$1.60
Tacoma to Bellingham	$3.10	Tacoma to Vancouver, B.C.	$5.10
Tacoma to Blaine	$3.75	Tacoma to Vancouver and return	$7.55

NORTHERN PACIFIC RAILWAY CO.

Rates via Puget Sound Electric Railway and Northern Pacific Railway:

Tacoma to Snohomish	$1.75	Tacoma to Vancouver, B. C.	$5.10
Tacoma to Bellingham	$3.10	Tacoma to Vancouver and Return	$7.55

CANADIAN PACIFIC RAILWAY CO.

Rates via Puget Sound Electric Ry. and Steamer "Princess Victoria":

Tacoma to Victoria ....$2.60; round trip, $4.50
Tacoma to Vancouver, via Victoria............ 5.10; round trip, 7.50

PUGET SOUND NAVIGATION CO.

Rates via Puget Sound Electric Railway and Steamer:

	Single Fare.	Round Trip.
Tacoma to Bellingham	$1.60	$2.75
Tacoma to Port Townsend	1.85	3.00
Tacoma to Port Angeles	2.60	4.50
Tacoma to San Juan Island points	2.60	4.50

ALASKA STEAMSHIP CO.

Rates via Puget Sound Electric Railway and Steamer "Whatcom":

	Single Fare.	Round Trip.
Tacoma to Victoria	$2.60	$4.50

PACIFIC COAST STEAMSHIP CO.

	Single Fare.	Round Trip.
Tacoma to Vancouver, B. C.	3.60	6.00

Passengers may secure tickets from the Alaska Steamship Co. and the Pacific Coast Steamship Co. via the Puget Sound Electric Ry., and have baggage checked directly to and from all Southeastern Alaska points.

Ask for Tickets via Puget Sound Electric Railway

IMPORTANT TO SHIPPERS.

We make very low rates and give *quick* Express Freight service on all *classes* of Freight from Seattle and Tacoma to all points on the Puget Sound Electric Ry., making connections with the Tacoma Railway and Power Co. in Tacoma for Point Defiance, Smelter, Steilacoom, South Tacoma, Spanaway, Fern Hill and Puyallup.

Also close connections in Seattle with the Seattle Electric Co. for Ballard, Green Lake, University, Fremont, Interbay, Latona, Ravenna, Madison St. and Madrona Park and *all* other Lake Washington points.

SPECIAL CARS AND SPECIAL TRAINS.

Low rates will be made for Special Cars attached to our Regular Trains; and Special Trains may be arranged for to leave at any time, day or night. This feature is specially recommended to Theatrical Companies, Clubs, Societies, Lodges, etc., visiting between Seattle and Tacoma. Train service may be arranged to suit their convenience.

For rates and arrangements, apply to Commercial Agent.

Spanaway Park, on T. R. & P.

Steilacoom Beach—the Village in the distance.

P. S. E. Ry. Limited, composed of motor, baggage, smoking and parlor cars, at speed of sixty miles per hour.

In White River Valley, near Renton Junction. Large dairies and fruit farms. Rich soil never failing crops. Rapidly settling.

P. S. E. Ry. track passing wagon bridge over Duwamish River, near Riverton Station. Surroundings most desirable for suburban homes, small berry and truck gardens.

"The Death of the Forest." Extensive timber lands, near Edgewood. This town is rapidly growing, the location being most desirable for small vegetable gardens and orchards.

Residences in Kent, a growing and prosperous little city about midway between Tacoma and Seattle. Surrounded by large dairy interests.

Famous Renton Coal Fields.

Wright Park in the Center of Tacoma, Surrounded by Many Beautiful Residences.

Bird's Eye View of Seattle.

Old Hudson Bay House and a picturesque oak, near Spanaway Lake, surrounded by beautiful picnic grounds.

This early day stage was PSE's first attempt at starting a feeder system in conjunction with the interurban. The first route was between Buckley and Enumclaw connecting with the interurban at Auburn. The company planned to extend it if it proved profitable. A similar feeder was tried between Seattle Heights and Edmonds on the Seattle-Everett Interurban but was closed after a short time for lack of patrons. (Author's collection)

and sailor club in downtown Seattle. The seven story club, financed by the traction company, had all the comforts of home. It could accommodate 350 men with another 500 beds in the old court house annex.

In May 1915, the company entered the auto-stage business with a route between Auburn, Enumclaw and Buckley, using three twelve passenger Studebakers, purchased for $4,500 each. This new venture was called Washington Auto Bus Company and if it proved profitable as a feeder to the Interurban, the company would consider similar systems elsewhere. Actually another connecting stage line had begun between Edmonds and Seattle Heights on the Everett Line but after several months was discontinued due to poor patronage.

In the meantime, paving of local highways and roads was continuing. King County awarded the largest contract to date to pave the highway from the 14th South Bridge in Seattle to Des Moines. The new highway, paved with local bricks, would shorten the distance between Seattle and Tacoma by seven miles and benefit users of the new auto ferry between Des Moines and Vashon Island.

In April 1917, 3.75 miles of highway between Duwamish and Renton Junction were also slated to be paved with concrete. Later, it was reported Tukwila citizens were holding up highway paving by obtaining a temporary injunction on the basis of promises made that brick would be used. However, later photos show the highway through Tukwila to be of Macadam paving.

In March of 1919, the East Valley Highway between Renton and

Auburn was scheduled to be paved, while King and Pierce Counties were at odds over the paving of the Hi-Line road between DesMoines and Tacoma. Yet the author remembers this stretch as remaining unpaved for many years.

The Enumclaw Co-Op Union Stage Line, making fifty minute trips to Seattle with connections in Auburn, opened a new waiting station in Auburn in early October. By December, two stage lines were in competition with the interurban between Auburn and Seattle. One operated out of the Union Stage Depot, and the other from Williams Auto Livery. A large ad in the Auburn weekly proclaimed, "Thirty stages each day to Seattle, Tacoma, Enumclaw and White River. Sixty cents to Seattle and forty cents to Tacoma, Auburn Stage Depot".

The Horseshoe Union Stage Line ran ads featuring their schedule in the same paper. They were identical in size to the ads placed by PSE and positioned directly above the interurban ad.

All this was slowly but surely eating into the Interurban's business. And, as if this weren't enough, Herbert Munter, an Auburn pilot, began a combination airplane-bus business between Auburn and Seattle in January 1920. The airplanes operated from nearby Munter Field in connection with buses into Auburn. Auburn was to be the headquarters for the new company.

On a hot day in June, for some unexplained reason, bricks on a section of newly paved Des Moines Highway exploded, wrecking a truck but causing no injuries.

In August 1920, the East Valley Highway opened for traffic to Earlington, bypassing Renton. Renton lodged a protest as the city thought it should come into Renton and connect with the new Sunset Highway.

In Seattle, there were 49 stage owners with 17 stage lines. The Interurban, faced with shrinking patronage, increased their schedules and were changing them frequently to meet the competition.

The end of the year brought protests from PSE over fare reductions by competing stage lines. Later the company decided to abandon the Puyallup Short Line because of declining business and stage competition. The rails were removed and the bridge over the Puyallup was dismantled and freighted to Duwamish to replace the aging wooden swing bridge there.

On June 15, 1924, PSE began a new half hour schedule on the main line between the hours of 9 AM and 7 PM. The half hour schedule only applied to trains between Seattle, Kent, Auburn and Tacoma. No change was made between intermediate points. To cover this increased service, the company rebuilt five parlor cars into four motor coaches. Evidently riders enjoyed the expanded service as business rose considerably. It goes without saying, but the company was doing everything necessary to win back business lost to competing stages.

In March 1925, George Standring opened a motor stage plant in Auburn. The stages were to be 30 to 36 passenger Pierce Arrows. Standring

Local train No.35 just west of Willow Junction in 1914. The depot can be see just behind the car. This photograph and the picture on the right of car No.525 passing Pacific City depot, is a series of photographs taken by the General Signal Company after the installation of a semaphore signaling system. The "44" on the signal post signifys 4.4 miles from Tacoma. Willow Junction ceased to exist after the Puyallup Short Line closed in 1922. Today, this would be the approximate location of a United Parcel facility in Fife. Opposite above, a northbound train is approaching semaphore 266, beside the Duwamish River in Tukwila. (James P. Graves collection)

was interested in providing new stages for a South Tacoma line and a new route between Seattle and Portland; Portland and San Francisco. By April 1925, Standring had the first 30 passenger, six wheel stage ready for the Auburn-Seattle run.

The new half hour schedule offered by the interurban, with minor changes from time to time, was to stay in effect up until the day of abandonment. To meet the new interurban half hour schedule, the stage line was now offering half hour service as well.

In June 1925, the state announced plans for a new state highway bridge over the Puyallup River at Tacoma. The bids for the new bridge were not released until September. The new bridge and viaduct were to be about 100 feet north of the existing PSE trestle and bridge, making it necessary to condemn and remove PSE tracks and the trestle to the tide flats. PSE tracks into Bay Street Yard also had to be relocated.

In May 1926, the Enumclaw Stage Company became Park Transportation Co. Five months later PSE took control of Park as well as the stages from Seattle to Auburn and Auburn to Tacoma.

In late October, Stone & Webster interests announced plans to combine several local companies into a new corporation known as The North Coast Transportation Company. This would include Park Transportation Company, the Tacoma Bus Company, the Portland-Seattle Stage Company and the stage division of PSE, but omitting the Interurban. Charles E. Parks was to continue as manager of the Auburn-Enumclaw Stage Line.

The new company, better known as North Coast Lines, was officially formed in January 1927, combining the before mentioned stage companies. In 1930, following the abandonment of Pacific Northwest Traction's interurban operations between Mount Vernon and Bellingham, the PNT stage lines north from Seattle and the Seattle-Everett Interurban, came under NCTC control.

The new highway bridge and viaduct over the railroad tracks and Puyallup River was opened with celebration in early January 1927. The new viaduct and bridge rendered a good portion of the Bay Street yards ineffective. The new highway, when finished in 1928, would eliminate four major railroad crossings and would shorten the distance between Tacoma and Seattle by nine miles.

In April, six new observation type stages were placed in service between Seattle, Tacoma and Portland. Four carried 23 passengers and were equipped with drinking fountains. They were built by Fageol and the Yellow Coach Company. 30 seat coaches, built by Yellow Coach Company (a subsidiary of General Motors), were in service between Seattle and Auburn while three new 36 seat coaches were on the Seattle-Tacoma run.

The State Highway Department awarded a contract in May for

This carbarn at 13th and A Streets in Tacoma was shared by both Tacoma Railway & Power and Puget Sound Electric. It was also PSE's headquarters until they were moved to Kent in 1919. In the barn, besides several streetcars, are PSE cars 529 and 523. This building was torn down sometime after the 1950's. (Lawton Gowey collection)

construction of 10.7 miles of 20 foot wide pavement from Tacoma to a point north of the road to Redondo Beach. This would open in the fall of 1927, leaving only three miles of unpaved road on the Hi-Line to Seattle. A new bridge over the Duwamish, south of Seattle, was scheduled to be built later in the year.

In June, several ads appeared in the *Auburn Globe Republican* for the Portland-Tacoma-Seattle and Seattle-Kent-Auburn-Enumclaw stage lines. Later that month, an NCL ad appeared heralding the new Fageol twin stages.

A new stage terminal opened at 8th and Stewart St. in Seattle in September 1927. The new terminal was for all stages and buses operating in and out of Seattle, as well as the interurban to Everett.

That same month thirteen miles of the new high line road (Highway 99) were scheduled to be paved between the Puyallup River bridge in Tacoma to a point above Des Moines. The remaining section to the Duwamish River would be paved in 1928.

Later that year, North Coast Lines advertised in local papers about traveling over the new Pacific Highway. The handwriting was on the wall for the Seattle-Tacoma Interurban.

A company photo of motorcar 516 and train near the company carbarn and shops. Originally parlor car 521, it was rebuilt and renumbered in 1912. In 1925 it was heavily damaged in a head-on collision with car 520, south of Kent. It was repaired and returned to service, lasting until the end in 1928. (Harold Hill)

A 1922 view of Ist Avenue South looking north from Spokane Street. No.26 West Queen Anne car has just crossed over onto the northbound tracks after a few minutes layover at the southern end of its route. (Author's collection)

Looking south on Occidental as a two-car PSE train loads at the Seattle depot. A close-up glass reveals the touring car on the right is sporting a 1919 license plate and the newspaper cart selling out-of-state papers, including the New York Times. A wagon and team belonging to James F. Keenan Grocery Company is under the Chauncey Wright restaurant sign. (Museum of History and Industry)

"Before and after" views of car 504 at the carbarns in Kent. One of the original cars, built by Brill in 1902, it was rebuilt in 1914 with parts from a similar car. The photo at left shows the car just after rebuilding. The circular emblem on the side is the Stone & Webster logo. Although PSE cars were usually single-enders, they all carried two trolley poles. The next photo shows a rather forlorn 504 at Kent just after abandonment. The strap across the rear door prevented trainmen from falling off the car when raising and lowering the trolley pole. (Both: Harold Hill collection). At bottom, a rear view of arch-windowed combine No.514 at the Kent barn. (Lawton Gowey collection)

Several views of the Tacoma depot before its remodeling in 1923. In the photo at the top of the page, the tower of the Tacoma city hall looms in the background, while above the four story building on the right, is the spire of the Northern Pacific headquarters building. Below, a Puyallup car is on the left with PSE car No.508 on the storage track behind the depot on the right. (Both: Washington State Historical Society)

Above, the information on the back of this photo says it is of the newly remodeled Tacoma depot in 1923, yet except for some new paint and trim very little seems to have changed. That's car 525 in front of the depot. (Harold Hill collection) At left, TR&P No.55 is laying over between runs at 8th and A Streets in Tacoma. On the right, doing the same is PSE No. 529. (Boland photo, courtesy Tacoma Public Library)

PSE No.600 is inbound to Seattle through Argo in July 1921. The streetcar in the distance is en route to Georgetown and South Park. (R.A. Hansen collection). At right, No.559 and the remodeled 502 pause between runs in the Massachusetts Street carbarn in Seattle. (Lawton Gowey collection)

THE RENTON BRANCH

From the beginning, the line to Renton was an important part of Puget Sound Electric's operation. The 14 mile branch split off the main line at Renton Junction, about 11 miles south of Seattle. Service was hourly with as many as 33 trains daily. In the beginning streetcars like No.557 and its trailer were used on the Renton line. Most of these cars came second hand from Seattle and Tacoma. In 1918 PSE obtained nine cars from TR&P, rebuilding them for use on the branch. They were numbered 600 to 608 respectively. In later years, larger cars such as No.558, 559 and 561 were also used in Renton branch service. The branch was also a major source of freight revenue for the railway. The company owned coal mine was located there as well as a brick making plant and a glass works that also provided additional carloads of freight to the interurban.

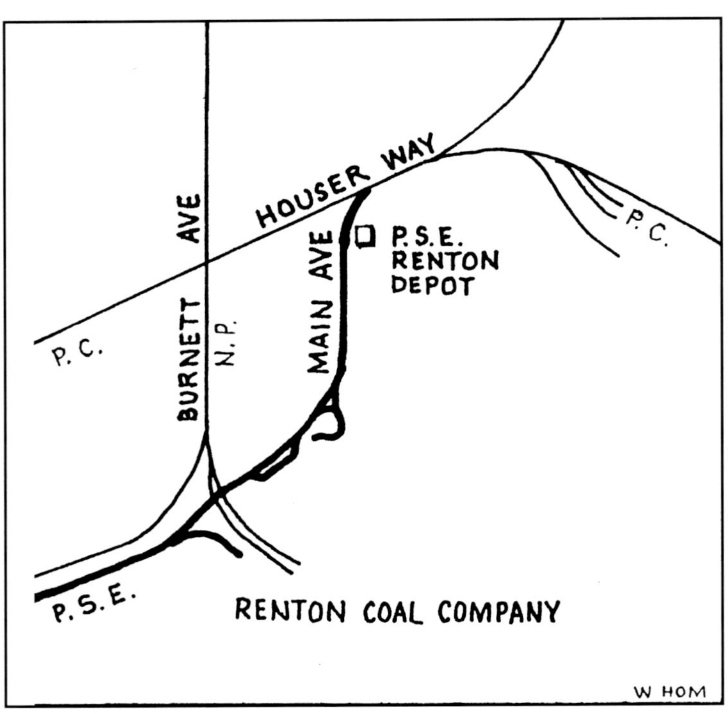

Below, train No.117 makes an early afternoon stop at the Junction while on the way to Renton. The depot is just to the left of the crew and passenger. Later, it was moved from inside the "wye" to a location near the site of the little shack in the background. (Lawton Gowey collection)

Above, looking north past the original depot at Renton Junction in 1912. At left, a year later a new Pacific Highway bridge is being built near the Junction. Today, this would be looking toward the north-bound lanes of West Valley-Interurban Avenue at Grady Way, one of the busiest intersections in south King County. (Both; Author's collection)

A Renton local ,with car 553 leading, is stopped for passengers at Quarry Station. (courtesy of Frank Cummings) Just below, Renton line cars No.556 and 555 pose on the high trestle at Renton Junction in the teen's. (Author's collection) Below, looking west on 3rd in Renton. The two tracks in the foreground are PSE's to the Denny-Renton clay (brick) plant. Beyond is the Pacific Coast-Milwaukee Road tracks in Walla Walla Avenue (Houser Way). The tracks on 3rd belong to the Seattle, Renton & Southern, a trolley line that ran from Renton along the west shore of Lake Washington, through Rainier Valley to Seattle. The square brick building is a Puget Power sub station (James P. Graves collection)

Above, an example of a 600 series car in its original form while under lease to PSE from Tacoma Railway & Power. The train is stopped for publicity purposes on 1st Avenue south and Massachusetts Street in Seattle in 1908. The train is getting ready to haul fans to Meadows race track. This was one of nine cars bought by PSE in 1918 and rebuilt for Renton branch service. (Lawton Gowey collection) At left, No.604 at the end of the line at 3rd and Main in Renton about 1919. (Harold Hill collection). Below, a pair of 600's laying over between rush hours at the Massachusetts carbarn (Author's collection)

At right, a three car train of 600's is entering the south Seattle carline at Argo (Airport Way and Lucile Street). The ivy covered building on the right is the Argo switching tower. (Author's collection) Below, train No.558, a Renton Line regular for many years, is crossing the Duwamish River, outbound for Renton in the 1920's. (Harold Hill collection) Opposite above, looking south to the swing bridge over the Green River at the Junction in 1912. (Author's collection). Opposite below, No.558 again at the Seattle terminal. As a single ended car it had to back from the Junction into Renton on the first trip of the day from Kent. The same occurred on the last trip, only in reverse. (Harold Hill collection)

Looking south toward Renton Junction

Here are several views of the "new" station at Renton Junction in its new location, away from the "wye". In the smaller view the new Pacific Highway bridge can be seen just to the right of the depot.(Harold Hill collection) The larger view, taken by conductor John C. White has train No. 107 pausing on its way to Renton in 1921.

Above, looking east to Renton from the end of the Junction trestle. In the background is the slag pile from the Renton coal mine. Present day Grady Way would be to the right of the track. (Lawton Gowey collection) Below, another John White photograph. This view of car 556 is at the Yesler Way-Occidental turn around in Seattle. This was the last car used on the Renton branch.

As the use of the automobile grew so did grade crossing accidents. In the early 1920's the company sent famed Seattle photographer Asahel Curtis to photograph some of the more dangerous crossings. This one depicts the crossing of South 180th at Orillia. In the foreground are the tracks of the Milwaukee-Union Pacific with the PSE's freight shed just behind. Out of sight behind the shed is the Orillia depot. A women is standing alongside the signal passengers used to let the motorman know they wanted to board the train. South 180th is also known as SW 43rd and is now in Renton. Orillia lost its post office in 1950. (Washington State Historical Society)

4. Accidents And Other Calamities

Over the years, the Interurban was plagued with accidents of every conceivable description; from head-on collisions and grade crossing accidents to people and animals getting in the way of the trains.

The earliest animal/train incident occurred even before the line opened. On July 31, 1902 a dog chased a cow, owned by a farmer named Gessner, on to the tracks near Georgetown where it was hit by Grant Street car No.20. Mrs Gessner complained the crew was too busy looking at their new barn to notice the cow in time to stop. A year later a train collided with a drove of cattle south of Georgetown. Six of the fattest cows were killed and local residents picked up enough free beef to last all winter.

An accident on December 4, 1902 wasn't so funny. Two trains, one from Seattle the other from Tacoma, collided head on at Summit (Edgewood). There were 80 passengers and crew aboard the two trains and while there were injuries, luckily no one was killed. Both lead motor cars were badly damaged and out of service for some time, necessitating the reduction of service to a train every two hours. Newspapers reported all the injured were on the train traveling from Tacoma. The wreck took place at the north end of Jovita on a curve near the switch by the sawmill. Each train was traveling at a fair rate of speed with the result that the Seattle car telescoped over the Tacoma motor causing the entire line to be tied up for more than an hour.

Two accidents marred operations a few years later in 1906. One rather minor, the second resulted in the line's first fatality.

The first occurred on the afternoon of December 6th, when the 5:50 train from Seattle jumped the rails while leaving the double track at Meredith Station. As it derailed, the lead car hit a pole, causing it to snap in two and fall onto the roof of the car. Damage was minor, not even a broken window. Passengers were transferred around the derailment until the line could be cleared.

Later that month, the company was not so fortunate. On December 26th, Train No.3, southbound to Tacoma, collided with a gravel train in a deep cut south of Edgewood, killing the conductor, motorman and one passenger.

Train No.3, consisting of a combination baggage & smoking section motorcar and two trailers, left Seattle at 6:30 AM. It was scheduled to meet No.6, the train from Tacoma, at Milton. But because it was running twenty minutes late, the meet was changed to Edgewood. No.3 took the siding at Edgewood to let No.6 pass. Following No.6 was a gravel train being pushed up the hill by a freight motor. The brakeman of the gravel train was riding on No.6 and had been told to inform the conductor of No.3 that the gravel train was following. However, it was never known if the order was delivered since the brakeman disappeared after the accident. Nevertheless No.3 left the siding, continuing down the grade at a speed estimated to be 25 mph. The work train was traveling about the same speed with the result the crash was heard over two miles away at Milton. The gravel cars tore into and landed on top of No.3's lead car with such force that it left its trucks. Two men, Conductor Ross and one passenger were killed instantly while the injured motorman, Guyer, died the next day.

Twenty-one others were injured but none fatally. Brakeman Griffith, of Train No.3, took charge of the rescue work and the wreckage was cleared by 7 PM with the injured taken to a Tacoma hospital by a relief train. Since the conductor died and the brakeman had disappeared, it was never known if the message was delivered, although one passenger said he had heard the order passed on. It was also never discovered why the brakeman disappeared. The site, known for years as "Deadman's Cut" was only recently filled to make way for a road to a new housing development. The remainder of the former line down alongside Hylebos Creek to Milton is now a sewer trunk line.

Eight years later, in October 1914, six cars of an interurban freight were wrecked when the long trestle over the Green River south of Kent, gave way from the train's weight. Motorman Davidson and his freight locomotive made it across safely while Conductor Guyer and two brakemen in the caboose remained on the far side. Two cars of paving bricks destined for the new Pacific Highway crashed into the river. Passengers were transferred around the wreck by bus until a temporary bridge could be built.

The following year theft of copper connections used in the new block signal system, caused a head-on collision just north of Quarry Station between Renton train No.111 and an extra freight. Both trains were traveling slowly and the only injuries were from flying glass. Horace Rumery was conductor on the passenger and A. J. Ecklund the motorman.

A damaging accident occurred at Willow Juncton on October 3, 1916 when a switch was left open by the crew of a Puyallup Short Line train. Limited No.20 left Tacoma at 4:35 PM and, according to investigators from the State Public Service Commission, was traveling at least 30 mph at the time of the accident. The lead car ran the switch, left the rails and overturned, taking a small coal shed with it. Thirty of the thirty-six passengers were riding in the trailer. Some of the injured were taken to a Tacoma hospital while the rest were sent on their way on the next train.

As the use of the automobile grew so did grade crossing accidents. On July 7, 1912, a grisly accident occurred at Riverton road crossing. A southbound train, traveling at 35 mph, hit a stalled auto belonging to Dr. E. M. Rinninger of Riverton. He was riding in the front seat with his chauffeur, when the car stalled on the grade crossing. For some reason the doctor stood up as the train approached. He was thrown out of the vehicle and dragged sixty feet under the interurban before becoming disentangled and rolling down the embankment. The train itself traveled 300 feet before

Here are several views of the wreck on October 7, 1914 when six cars of an interurban freight wound up in the Green River, south of Kent, when the long trestle and bridge gave way from the weight. The motor car made it safely across with the caboose remaining on the far side. Four cars of brick intended for paving the Pacific Highway south of Auburn were included in the consist. The steel bridge in the background carried trains of the Milwaukee and UP. Much of the former PSE right of way is now a hiking trail with a foot bridge over the river next to the UP steel bridge. (Both: Author's collection)

P. S. E. RY.
FRYE & COMPAN
505

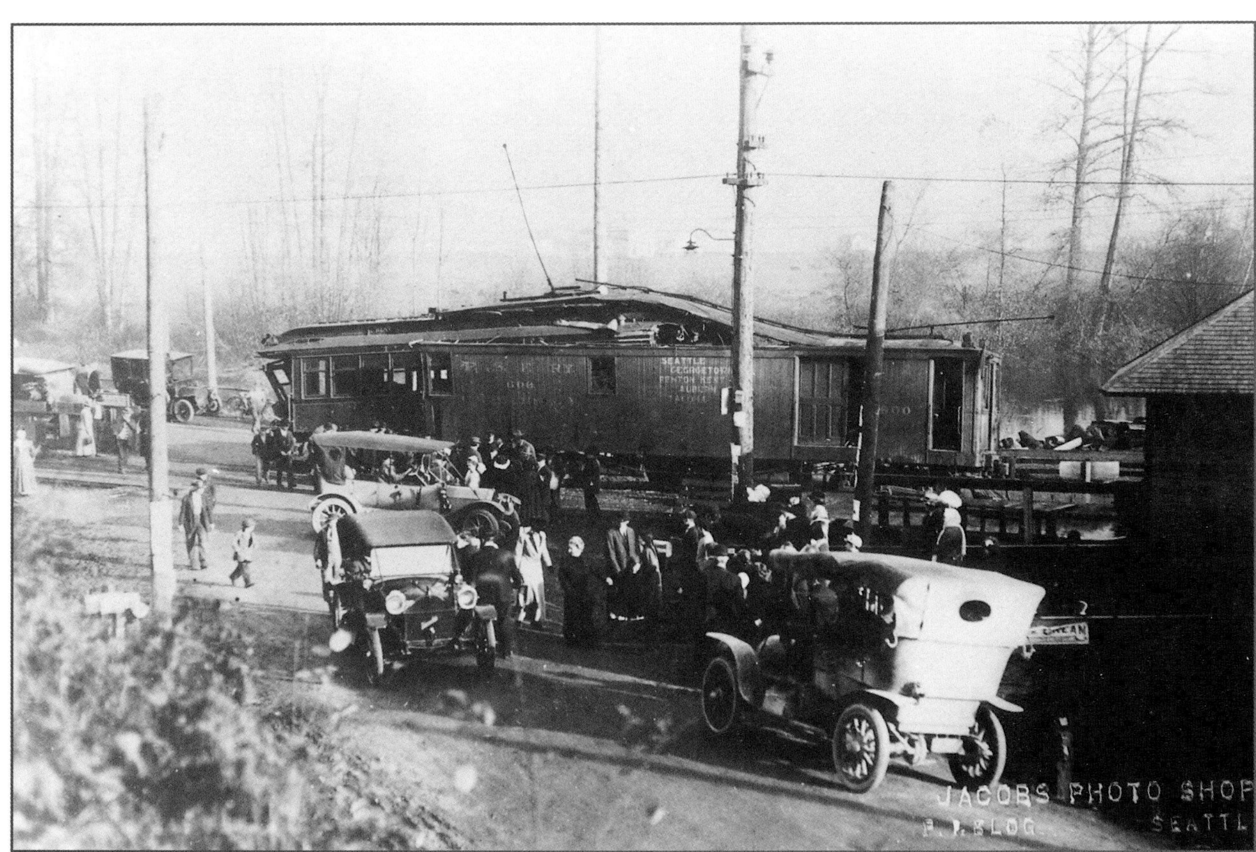

The road crossing at Riverton was the scene of several bad accidents. The one above occurred in November 1912 when a freight motor, following too close in heavy fog, rammed a standing passenger train.(Author's collection) Car No. 511 was severely damaged in the crash. It was stored at the Kent shops until World War I, when it was rebuilt and put back in service.(Harold Hill collection) Opposite above right, after this accident the company employed a guard at the crossing (Tukwila Historical Society) At right, Riverton was also the scene of this January 1914 slide. Today, this would be 42nd Avenue South and Interurban Avenue in Tukwila. The wooden drawbridge over the Duwamish would be South 124th. Most of the hillside has given way to industry and Metro's south base. (Washington State Historical Society)

coming to a stop. When it did, the crew ran back to the scene of the accident. Another physican, a Doctor Fleisher, heard the crash from his home a half block away and rushed to the scene. Two women passengers in the auto were taken to nearby Rosenberg's store, where they were given first aid by Dr. Fleisher. A photo and an account of the accident appeared in the next morning's edition of the *Seattle Post Intelligencer.*

This limited train ran through an open switch at Willow Junction and overturned, smashing a small shed and injuring thirty passengers. This accident, on October 3, 1916, was blamed on the crew of a preceding Short Line train for failing to close the switch. (Museum of History & Industry)

As a result of this accident and another that occurred several months later, when a train was hit from behind by a speeding freight motor, the Auto Club and State Public Service Commission made an inspection of PSE grade crossings. The crossing at Bluffs, southeast of Auburn, was determined to be the most dangerous in the state. A grade separation was deemed imperative. The Riverton crossing was slated to be improved by moving the highway west over the hill. This would later become East Marginal Way. A watchman was stationed there until the new highway was built. Another crossing, at Orillia, was determined to be unfeasible for grade separation.

Not long after this series of accidents, PSTL&P began running promotional ads in local newspapers extolling management, relationship with communities and service to its patrons.

At the town of Pacific, the crossing was especially dangerous with two serious accidents in as many years. In 1916 three men were killed when a northbound interurban slammed into their car at the Main Street crossing. An inquest, held a month later, placed the blame on the interurban company for having poor warnings at the crossing.

The second accident occurred in 1918, very similar to the first, when another northbound interurban hit an auto at the same crossing. It seemed the interurban company hadn't learned a thing from the earlier accident. Local residents complained the crossing signals were not loud enough and could be heard less than a block away.

In March the same year two people were killed in an auto-interurban collision at Van Asselt crossing near Georgetown. The driver apparently was trying to beat the southbound limited to the crossing.

Meredith was the scene of several bad accidents in October 1918. In the first, a woman made the mistake of stepping in front of a freight train. She obviously had mistaken the freight for a passenger due to stop at the station and was struck by the speeding train.

Later that month, while making its early morning rounds, a milk wagon was hit at the same location. In what sounds like a bad Keystone comedy, the driver was thrown through the interurban cab window escaping with minor cuts and bruises. The motorman was badly cut by flying glass yet was able to continue the run. The wagon was demolished and the horse, suffering a broken leg, had to be shot.

A few years later in August 1920, a Seattle family, returning to the city from an outing in their car, collided with an interurban at Edgewood crossing. The vehicle was going down hill on Military Road and because of

brush obscuring the right of way, the driver did not see the limited. All members of the family were injured and the driver killed.

1921 turned out to be a bad year for auto-interurban collisions. Four people were killed at Algona when the driver failed to stop for a warning signal and drove right into the path of a speeding limited train. A driver at Auburn was more fortunate in a collision at Main Street crossing. The interurban was slowing for the depot on the far side of the crossing and the motorman had shut off the power. He was coasting into the depot when the driver attempted to beat the interurban.

The same crew, involved in the Algona accidents, was involved in another at Meredith crossing, this one costing the life of a Japanese farmer.

Puget Sound Electric wasn't the only one involved in hitting autos or pedestrians; the Northern Pacific had its share of accidents too. An August 1921 accident involving an NP train and a Seattle-Auburn stage at Kent, severely injured five passengers.

On September 2, 1927, a limited interurban hit a truck at Boyd's Crossing, one mile south of Auburn. The limited was southbound to Tacoma when it hit the stalled vehicle. The driver was still trying to get the truck started when it was hit. The truck was carried 200 feet down the tracks before the electric car was able to stop. The mishap ruptured the truck's gasoline tank and a spark from the interurban's electrical system started a fire, engulfing the truck and spreading to the interurban's wooden frame. Motorman Hyre and Conductor Baker of Allentown, tried to put out the fire with a small extinguisher, but without success. By the time the

A "cornfield meet" occurred south of Kent depot on September 23, 1925. Two trains, northbound No.520 and southbound No.516 were scheduled to meet at Thomas but because No.516 was running late, the meet was changed to Kent. 516 was standing at the station as the conductor received his orders when a passenger pulled the bell cord signaling the train to proceed. The resulting crash wasn't as bad as it looked, both motormen survived and only thirty passengers were slightly injured. Car 516 was repaired and returned to service while 520 was shoved to back of the Kent barn, remaining there until abandonment. (Author's collection)

The depot at Thomas was typical of most PSE way stations. Each was equipped with a lever that raised a flag to let the motorman know he had passengers to pick up. This device was used only on locals, as limiteds were nonstop. While limited motormen would signal upon approaching a station, over the years several people were killed stepping in front of speeding trains thinking they were locals. One rainy day, a lady actually mistook a freight for a passenger local and was fatally injured. (Harold Hill collection)

Auburn Fire Department arrived it was too late, the interurban car had burned to its trucks. Even though the car was filled to capacity the passengers escaped unhurt. The mishap tied up interurban traffic for hours.

A head-on collision between two trains occurred south of the Kent depot on September 23, 1925. Southbound train No. 516 was standing at the station when a passenger inadvertently pulled the bell cord signaling the train to proceed. A Mrs Neiswender, riding in the second car of the train, realized something was wrong when it started to move while the conductor was still in the depot receiving his orders. She tried to reach the motorman but only got to the front of the train when the crash occurred. Motorman Davenport on northbound train No.520 saw the headlight of the 516, set his brakes and jumped. The motorman of the 516 rode out the crash and miraculously survived. While there were over fifty passengers on board the two trains, only eleven were slightly injured.

The interurban also had its share of natural disasters. In December 1902, the line was shut down briefly by several slides caused by heavy rains. Later that winter there were more problems due to flooding on both the main line and the Renton branch.

In December 1917, severe flooding again hit the valley causing a Tacoma bound interurban to jump the track near the switch south of Auburn Station. The cars were filled with soldiers returning to Camp Lewis and luckily, only one man was injured as the lead car tilted over on its side. The track was ripped up for a considerable distance causing the line to remain closed from 10:40 PM until 6:00 AM the next morning. The flood also shut down the Northern Pacific. In a related incident, a bus from Tacoma to Puyallup stalled in five feet of water just east of Puyallup. Water was also reported cascading down the streets in Kent.

1918 ushered in the third flood in as many weeks. In the White River Valley, damage occurred between Thomas and Orillia including Kent. Although the interurban escaped unscathed, both the Northern Pacific and Great Northern, north of Seattle, were shut down. There were several serious floods in 1921 yet the interurban was able to keep operating.

In mid-September, the Interurban was out of service for several hours due to a forest fire at Edgewood. Sparks from a logging locomotive were blamed for starting a blaze that burned the poles carrying high tension lines supplying power not only to the Interurban but to the street railway lines in Tacoma. The raging fire consumed one thousand cords of wood piled alongside the Interurban and destroyed most of the small town of Edgewood.

In early February 1916, the Puget Sound area was hit by a record snowfall. Seattle itself was particularly hard hit causing the city to be shut down for days. In a February 5th editorial, *The Seattle Star* called for the public to, "Grab a shovel folks! Let's give Seattle Electric a hand".

Early the following week, northbound train No.44 smashed into a

slide three quarters of a mile south of Stewart's Point tunnel, causing the lead car's pilot and fender to be carried away. The train was late due to heavy snow. Fortunately there was a maintenance crew nearby who cleared the track, allowing the train to proceed to Auburn. An extra train was sent from Seattle to pick up the passengers and take them on to Kent and Seattle.

Lastly, not all mishaps involved gore, broken bones or fatalities. One funny incident occurred during the early 1920's when an eleven year old boy, leaving his aunt's home in Seattle for his parent's home in Auburn, rode an interurban cow catcher from downtown Seattle to Renton Junction before being discovered.

Here are several more photographs from the series by Asahel Curtis taken in the early 1920's. This is the Military Road crossing at Edgewood and considered one of the more dangerous. The photo opposite gives an overall view of the crossing. The picture at the top of this page looks north (actually southeast). In the distance is semaphore No.90 and the Edgewood passing track. Below, the camera has swung around and is facing south. In August 1920 a car carrying a Seattle family collided with a limited train here killing the driver. Brush along the right-of-way obscuring the drivers vision, was the reason given for the accident. This area is still pretty rural today although Military Road has been raised about ten feet. (All: Washington State Historical Society)

Until the 1950's, when the Howard Hanson dam was built on the Green River, the White River Valley was subject to severe flooding from time to time. At left are several scenes from the flood of 1906, just south of Kent. At right two pictures of Kent during a period of high water. The photo opposite above looks north to the Kent carbarn and depot from Titus Street. (Boland photo, Tacoma Public Library) The photo opposite below is looking south. Flooding caused the Interurban to be shut down a number of times. Both of these photos were taken before the Milwaukee Road was built. (Clark of Kent, Author's collection)

Interurban Freight

Puget Sound Electric freight equipment consisted of two wooden box motors, three center cab locomotives, 22 boxcars, 49 flat cars, 35 hoppers, 34 gondolas and three cabooses. The equipment was used for a variety of hauling that included express, brick and coal from Renton, produce from the White and Puyallup River valleys and milk from farms and dairies along the line. There were special meat (refrigerator) cars from Carsten's in Tacoma and from Frye Packing Plant in Seattle. Newspapers were delivered at various stops by passenger trains.

As the area developed, building material was moved by interurban rails.This included most of the brick used in paving the Pacific Highway through the west side of the valley.

The Puget Sound Electric maintained freight houses in both Seattle and Tacoma and smaller freight sheds at nearly every station along the line. Log trains were numerous to the mill at Milton. The company also had its own gravel pits at Renton and Edgewood.

Seattle and Tacoma street railway systems were actually extensions of PSE as both systems were owned by the same parent company. In Seattle, PSE turned over their freight cars to both Seattle Electric and Pacific Northwest Traction at Massachusetts Street. In Tacoma it was Tacoma Railway and Power and Pacific Traction that distributed freight originating on PSE.

The line had a "Spud Local" which ran daily right up to the end. The name derived from carloads of potatoes it handled from the valley farms to the markets in Seattle and Tacoma.

Coal from the company mine at Renton was one of the leading producers of freight revenue until it petered out. Most of the old mine property is still owned today by Puget Sound Power and Light

The interurban had competition from four steam roads at nearly every community along its line. Thirty percent of the company's business was freight, yet as roads were built and others improved and paved, trucks began siphoning off the remaining freight business.

The company tried to meet the stiff competition by reducing rates, and starting an express delivery system in Tacoma and Seattle. It was to no avail, by the 1920's the freight business had declined until there was practically nothing left.

PSE No.601 heads the "Spud Local" through Tukwila about 1915. Always popular and profitable, the train lasted until the end of interurban operations. Motor 601 was renumbered 1601 in 1918 when the TR&P Jewett interurbans were rebuilt and renumbered 600 to 608 for Renton Line service. (Author's collection)

At left, brand new express car No. 552 poses next to the brand new Kent shops about 1903. It was renumbered 600 in 1907 and 1601 in 1918. The car was scrapped in 1926. (Clark of Kent, Author's collection) Below, freight motor No. 626, in its original form, heads a string of homemade flatcars on 1st Avenue South en route to Meadows race track in 1906. The race track would be what is today the south end of Boeing field. Trains to the track operated in the summer months only. The slats and roofs on the flatcars were stored at Kent during the winter months. The 626 sits by the Massachusetts Street carbarn after rebuilding in 1914. Brakeman Al Rockwell stands at right. After the Interurban quit, Rockwell and his wife moved to Tacoma where he went to work on the streetcars. When they were replaced by busses in 1938 he became a driver, retiring after World War II. (Both: Author's collection)

At the top of the page, ex-PSE No. 1626 in service on the Tacoma Municipal Railway near the tide flats in 1935. (Emery Roberts) Below, the 1626 wound up on the Skagit River Railroad, operating out of Newhalem, Washington. The locomotive is shown there on September 9, 1953. The cars are from the Oregon Electric, two of which became part of Andy's Diner on 4th Avenue South in Seattle. (Lawton Gowey)

Freight motor No.627 was a homemade machine built by PSE in 1907. In the insert photo she's posing at Renton with a coal train. Motorman "Dad" Kelly is standing on the car to the right. His son Roy is on the ground. Roy left PSE to work for British Columbia Electric but returned in time to run the last PSE train out of Seattle. He later went on to become a successful restaurant owner in Renton. (Roy Kelly) Below, the 1627 is northbound with a trainload of bricks and coal through Tukwila in 1918. (Author's collection)

At right, the much rebuilt 1627 on the Tacoma tide flats in 1938. Notice the extremely thick underframe and long hoods. The original Baldwin trucks have also been replaced by Brill 27E 's.(Emery Roberts) Middle, 1627 went to the Skagit River R.R. in 1944. She is shown at Newhalem on March 6, 1954 just before the line was abandoned in favor of the North Cascades Highway. (Lawton Gowey) Below, the 1627 is being returned down the lift from the upper level at Newhalem. The railroad is gone but the lift is still in operation, used for tours given by Seattle City Light during Railroad Days celebrations (Harold Hill)

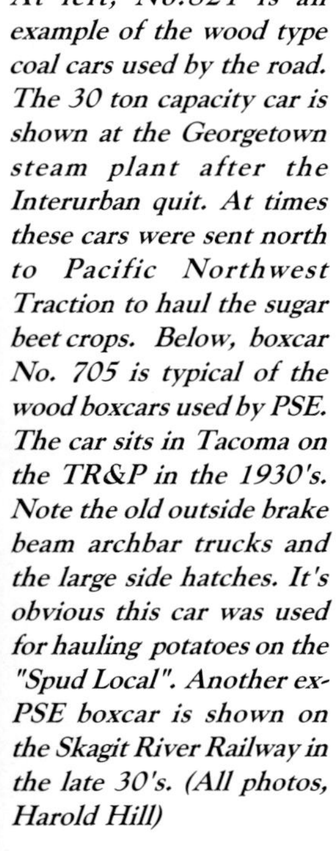
At left, No.821 is an example of the wood type coal cars used by the road. The 30 ton capacity car is shown at the Georgetown steam plant after the Interurban quit. At times these cars were sent north to Pacific Northwest Traction to haul the sugar beet crops. Below, boxcar No. 705 is typical of the wood boxcars used by PSE. The car sits in Tacoma on the TR&P in the 1930's. Note the old outside brake beam archbar trucks and the large side hatches. It's obvious this car was used for hauling potatoes on the "Spud Local". Another ex-PSE boxcar is shown on the Skagit River Railway in the late 30's. (All photos, Harold Hill)

PSE's Massachusetts Street carbarn and storage yard in Seattle sometime in the mid-teens. Presently this is the site of a City Light sub-station. (Author's collection)

By the look of the construction tents in the foreground, this photograph of PSE's freight depot on the Tacoma tide flats was taken soon after the line was built. With the decline in freight business it was gone by 1927. (Museum of History and Industry)

PSE caboose 001 was one of four similar cars built at the shops of the Tacoma Railway & Power Company around 1911. The 30' cars rolled on wood frame diamond archbar trucks and weighed 27,600 lbs each. This photograph reproduced from a copy of the "Electric Railway Journal" shows its original paint scheme, "the company's standard green". However the photo below, taken years later on Massachusetts Street in Seattle, shows one of the cabooses painted in what appears to be a light color. (Author's collection) Ex-PSE wrecker 1650 is pictured in Tacoma's Bay Street yards in 1938. It was built by the TR&P in 1910 and scrapped by Tacoma Muni in 1940. (Vernon Rutledge)

Looking north at Argo Crossing in the 1920's. This was the Author's favorite spot as a youngster to watch trains and trolleys. As many as forty passenger trains and numerous freights passed daily. In addition six interurbans and eight streetcars crossed at this site every hour. Here, PSE tracks crossed over the OWRR&N (UP) and left the South Seattle tracks for 1st Avenue. South Seattle cars crossed over the Pacific Coast-Milwaukee line to Stacy Street. There was the NP main line to King Street Station and the Milwaukee-UP tracks to Union Station. In the days of two man streetcar operation the conductor would walk ahead of the car as a flagman. The Olympic Foundry building on the right is still there. (Museum of History and Industry)

5. THE END

An ad appeared in the *Auburn Globe Republican* in January 1928 regarding North Coast Lines modern stages to Seattle and Enumclaw. It included schedules and in small print at the bottom were the words "And Interurban". In the same issue, was a heading about PSE's defaulting on $2,427,000 interest payments on company bonds that were due on February 1, 1928. The headline screamed: "ELECTRIC LINE HAS TROUBLES".

The article went on to say the company had also defaulted last August and had been losing money for some time. It had even failed to meet charges and operating costs. The parent company, Puget Sound Power & Light, had decided in January not to advance additional funds to its subsidiary, Puget Sound Electric. The article also noted that two other Puget Power subsidiaries, Tacoma Railway & Power and Pacific Traction were also in financial difficulty.

On February 23rd, on complaint of Old Colony Trust of Boston, attorney Scott Z. Henderson, was named receiver by the Federal Court in Tacoma. An injunction was promptly filed to prevent the company from selling any of its property.

T. M. Morgan Paving Company was awarded a $256,497 contract to pave the remaining section of the highline road from Redondo into Seattle. The new highway was billed "The Nation's Best-New Road and is Unequaled".

In August, several North Coast Line ads appeared in local newspapers; this time there was no mention of the interurban.

Three months later, a large ad in the Auburn weekly gave the connecting schedules between the interurban and the stages in Tacoma for Portland. There were seven daily connections. Also, the fares were reduced to $3.90 one way and $7.22 round trip.

A week later a headline in the *Auburn Globe-Republican* read: "MAY ABANDON TROLLEY LINE".

The newspaper went on to say that the Seattle-Tacoma electric line was in a bad way financially. The Federal Court in Tacoma would act on Saturday and Federal Judge, Cushman in Tacoma, has asked an order be prepared, calling for the closing of the line. His order followed an appearance by the receiver, Henderson, with an application for the Interurban's sale. It was disclosed at the hearing that the Interurban had been operating at a loss and that revenues had been steadily declining. L. L. Lamb, road supervisor for PSE, testified if the road was to continue to operate, it would be necessary to build a new bridge at Renton Junction at a cost of $24,000, and that other costs of improvements would amount to an excess of $17,000. He stated that there were no funds available for such improvements.

On October 18th, the final link of the new highway between Tacoma and Seattle was opened with dedication ceremonies taking place

PSE cars 523 and 503 pause for a station stop at Kent while heading north-bound in the late 1920's. (Harold Hill collection)

near Angle Lake. Governor Roland Hartley spoke as did the mayors of Seattle and Tacoma and auto club president, J. W. Maxwell.

With the annulment of PSE electric trains, Auburn was thought to be the hub of future stage activities. PSE, in the next two months, passed out questionnaires to their patrons concerning their traveling plans once the line closed.

North Coast Lines made it known more stages would be added to their existing schedules, effective January 1, 1929. A half hour schedule between Auburn and Seattle would be instituted. In a suit favorable to North Coast Lines, the State Department of Public Works issued a certificate of necessity for NCL to operate stages between Seattle and Tacoma, via Auburn and Pacific and over the Milton cutoff.

PSE filed notice to terminate interurban service as of midnight, December 30, 1928. Formal notice of annulment of all operating tariffs and schedules of PSE were filed in Olympia. Cancellation would be effective by an order of Edward E. Cushman, Federal District Judge in Tacoma.

On December 27th, a schedule of busses from Auburn to Seattle, effective on the closing of the interurban, appeared in the Auburn weekly newspaper.

The last train from Seattle to Renton made its run on the evening of Saturday, December 29th, with Howard Wellman, motorman, and Horace Rumery, conductor. The next night, Sunday December 30th, motorman Roy Kelly left Seattle with the last interurban to Tacoma. At

Tacoma he returned the train on schedule to Kent for the final time. The company sent Kelly and his crew out on the line to pick up and bring to Kent all remaining freight cars. Once they returned to Kent, in the early hours of December 31st, the power was shut off and the twenty-six year interurban line had come to an end.

In the following Thursday edition of the weekly, *Auburn Globe Republican,* appeared a headline and story:

"INTERURBAN IS NOW HISTORY, FAMOUS LINE QUITS OPERATION".

It went on to say:

"E. E. Wood, senior conductor, gave his signal to his motorman to depart from Auburn for the last time last Sunday night. Just as they had been doing for twenty-six years, two Seattle-Tacoma cars pulled into Auburn Station on Monday morning just a few minutes after midnight. A few passengers alighted and a few boarded and then they were off into a misty night, with red tail lights blinking, never to be seen again. But last night was not as gloomy as many of the big city newspapers had predicted. Both trains, the one from Seattle, and the one from Tacoma, a few minutes later, were filled and the cars were warm and cozy.

The two trains, one at 12:02 and the other at 12:06 AM, were the last that Auburn was to see of a rapid transit service that will go down in history as one of the best. Auto buses and private autos, free of heavy taxes, had seriously cut into patronage for several years. Rolling stock and equipment will be sold, shops and tracks to be removed.

Many old timers are loathe to believe it had been abandoned. They wanted to know why the company had laid new ties, put in new piling and improved the roadbed during the last few months, if they, indeed, intended to scrap the line."

The newspaper account continued,

"The first train had run on September 25, 1902, only eleven days after actual construction had been completed. The interurban had come as a God-send to valley residents. Travel before that had been by horse teams and wagons over impassable roads. For years the electric trains had served the farmers with milk trains that carried their output to Seattle and Tacoma markets. One of the line's freight trains to its closing days, was known as the Spud Daily, because of its original service to the potato growers of the valley.

The line was famous for years with tourists, as the big heavy cars, driven at high speed through a land of exceptional scenic beauty, made the ride a popular one.

Three and four car trains were required at times to handle the crowds with parlor cars on the rear at an extra fare. It was known as one of the fastest interurban lines in the nation and always ranked lowest in accidents. The fastest time ever recorded was made by a special train.

That train covered the thirty-six miles in fifty-one minutes, stopping only at Kent and Auburn en route. Local trains making stops at other stations, required eighty-five minutes.

Four employees, F. J. Dunn, F. H. Andrews, H. A. Wellman and D. M. Dingwall, have been with the company from the beginning. All older employees have been offered positions in other branches of Puget Sound Power and Light. Several are to become stage drivers with North Coast Lines. All told, there were 125 employees when the line shut down. The line itself is to be sold as junk."

On February 14, 1929, former motorman and now stage driver, C. S. Paine, was injured when his Auburn twin coach was forced into the path of a Seattle streetcar. Paine swerved his stage to avoid hitting a private auto. The stage had left Seattle at 9:55 PM and was on an icy street at the time. Paine suffered a broken leg and was cut by flying glass.

Only three offers were made at a public auction held in late February to buy the defunct PSE equipment and right of way. The highest offer was by PSP&L for $348,208, while Perry, Buston & Doan Company of Boston offered $251,878 and Hofius Steel & Equipment Company and Hyman-Michaels of Seattle made a joint bid of $178,250 for operating property. The bids were turned over to Judge Cushman, who will decide on March 11th if the bids are proper. A physical value of $389,541 was given by Henry L. Gray, appraisers.

The PSE sale was held up for one week as a new bid from San Francisco interests arrived at the eleventh hour, proposing $360,000; $12,000 higher than that made by PSP&L. The bid was made by representatives of William Rosenthal and Louis Rothenberg. Rothenberg was a member of a marine wrecking company and Rosenthal owned a brass and metal company.

The sale was held up for one more week on March 21st. Eventually the bids were thrown out by Judge Cushman due to legal entanglements. They were not resumed until 1930, and in the meantime opposition arose over the line's shutting down.

The Interurban Confederated Community Club was formed at a meeting of interested citizens at Fife City hall. There were one hundred or more citizens at the meeting to find ways to bring the Interurban back. The majority of those present were unhappy with the bus service that had replaced the electric cars. Some expressed the view that the Interurban had indeed made money, in spite of what the company had claimed. It was said the line had made a profit of $256,000 in 1928 and $257,000 in 1927. While there was much interest in reopening the line, ideas on how it should be done were lacking.

Scott Z. Henderson, receiver of the defunct PSE, appeared before a meeting of the Interurban Federated Clubs and claimed that PSE was making money when it quit, but maintenance costs, building new trestles and bridges made it wise to discontinue operations. No mention was made of the new highway which shortened the distance between Seattle and

Several stage lines made connections with the interurban cars at the 8th and A street station in Tacoma. Eventually , in January 1927, most of these lines were merged into one, North Coast Transportation Company. In a 1925 view at Tacoma station a stage from Olympia is in front of another bound for Steilacoom. In the insert photo , several Tacoma south end streetcar lines looped at 8th and A along with the Interurban. They included the McKinley Park line, Portland Avenue, South Tacoma and Pacific Avenue cars. (Both: Boland photo, Tacoma Public Library)

Tacoma by several miles. Henderson also said the Interurban could be purchased for $300,000, if financiers could be interested in its operation. It could be a paying proposition but he didn't feel PSP&L would be interested as they wanted to get out of the transportation business.

On March 20th, newspapers reported that "Interurbanites may organize permanently". A suggestion was made that the Interurban Federated Clubs meet for all kinds of activities. Meetings were slated for Jovita and Fife.

Again, on April 10, 1930, PSE property was slated to be sold to the highest bidder. A week later a date was set for the sale of the PSE property and equipment. All rolling stock and equipment would be auctioned off at a public sale to be held at Kent on June 16, 1930. No minimum amount was fixed. All bids were to be accompanied by a check for $10,000. The property would be sold free of all liens and encumbrances. This would be the final chapter of the "rise and decline" of the famous third rail electric line and the road will be finished forever.

The auction was held on June 16th and PSP&L was the highest bidder for the right of way and substations. They went for $213,750 while the rolling stock was sold to Pacific Equipment Company of Portland, Oregon for $135,000. When the sales were approved by the federal court, the receiver went along the line discouraging residents from trying to get service resumed.

Some communities were still hopeful that the distribution of property meant service would be resumed with lighter cars.

It was noted in the Auburn weekly newspaper that the first sale had been nullified by Judge Cushman because of a legal tangle between interested parties.

On September 25, 1930, twenty-eight years to the day since service had begun, a photo, taken by the *Tacoma Ledger,* appeared on the front pages of local newspapers showing rails being removed at Goldau Station. Goldau was just east of the Puyallup River.

In one newspaper, an accompanying article stated that the wrecking company had already removed the third rail and was now tearing up the old line from Tacoma to Seattle. The porcelain insulators that had cost the company $8 when new, would in all probability be discarded along the right of way.

Ever since the interurban had shut down, people along the line had protested and tried to get someone interested in resuming service, but to no avail. With the new highways, motor coaches, trucks and autos, it was simply too much of a risk. The Interurban Club organization that tried to find ways to resume service, had failed too.

Opposite, it's near the end of interurban service as car No. 512 and companion No.525 arrive at the Tacoma depot. Sometime later 525 was destroyed by fire near Jovita. (Harold Hill collection) Opposite below, It's December 30, 1928, the last day of service on the Puget Sound Electric. Conductor White has turned his camera over to a station employee and stands with the motorman for one last picture. This car will make three more trips before calling it quits. No information as to the final dispostion of this car has yet surfaced.

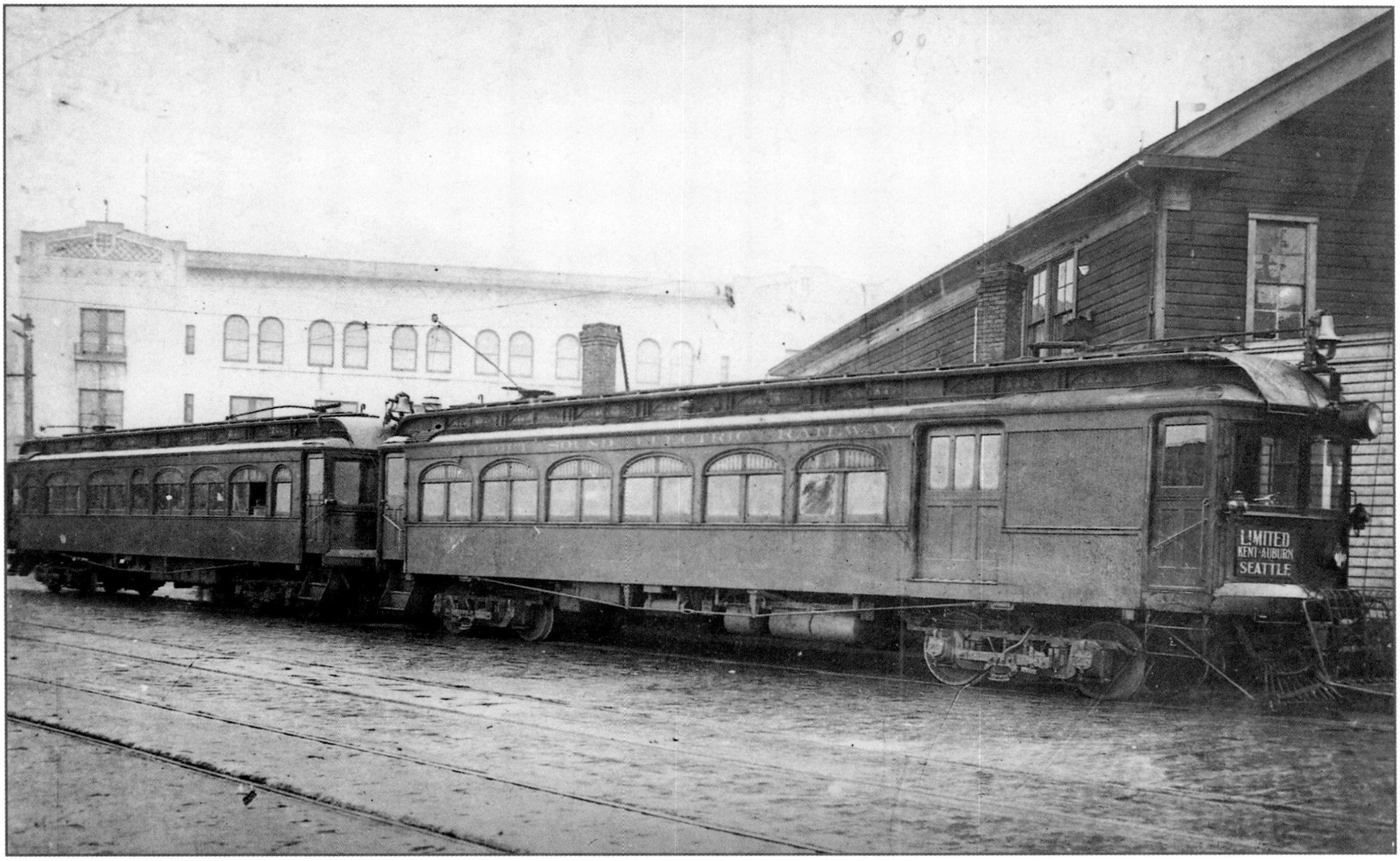
LIMITED
KENT-AUBURN
SEATTLE

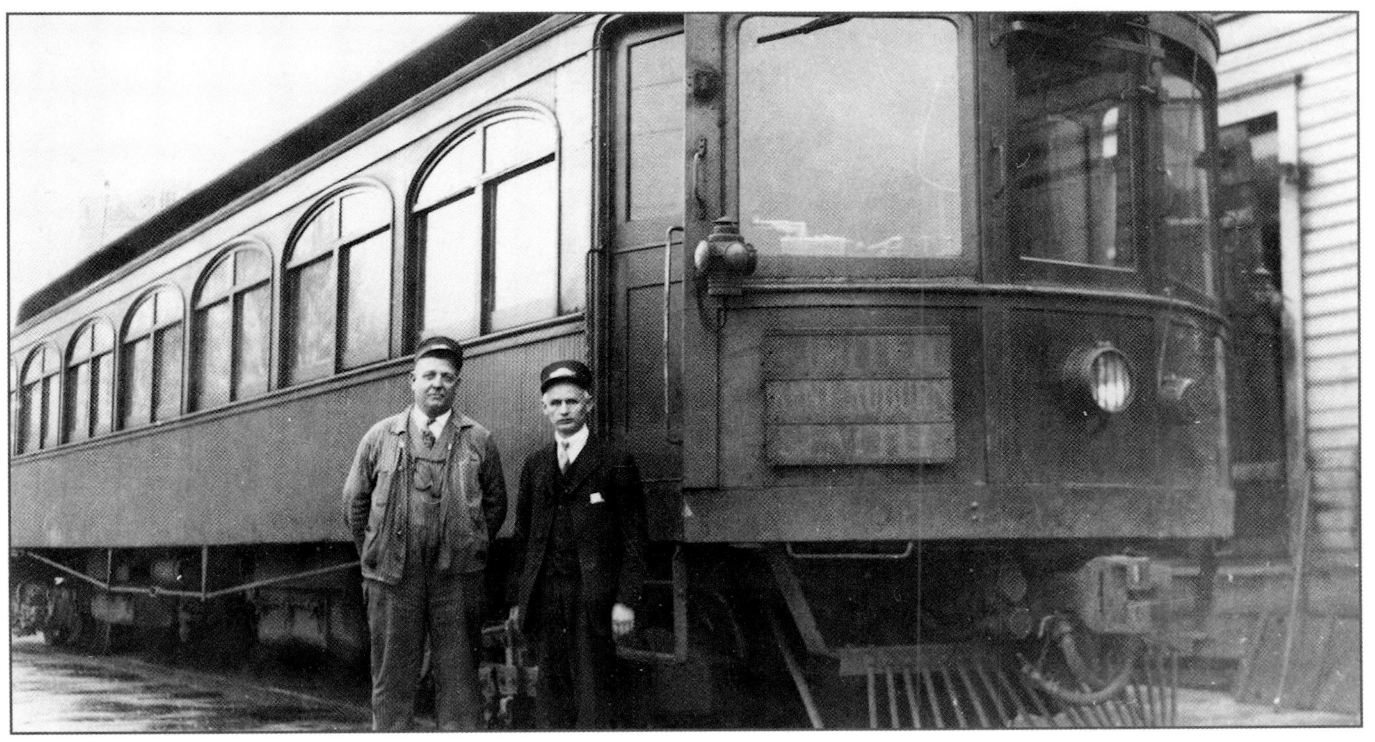

Today very little remains of the Puget Sound Electric...a few buildings and vestiges of the old right-of-way here and there. At the top of the page the old photo of Duwamish Avenue (Airport Way) in Georgetown, looking south, was taken in 1921. The tracks were used jointly by PSE and the city's South Seattle No.6 cars. (Lawton Gowey collection)The color photo shows the area today. Many of the old buildings remain and the street has been covered with asphalt. The photo below is a 1927 view east of Fife of the freshly graded roadway that will shortly become the new Pacific Highway. (Pemco Webster & Stevens collection, Museum of History and Industry) The color photo presents a striking contrast. Except for the house at the extreme right in both views, the area today is almost unrecognizable. At left, this photo, taken by the author in 1958, is the old right of way at the north end of the abandoned PSE tunnel high above West Valley highway near Jovita Boulevard. In the old photo at right, a baseball game is in progress at the Fred Nelsen farm at Renton Junction. Note the passengers watching the game from a Renton bound interurban. (Author's collection) The insert photo is the Nelsen barn as it appears today. Opposite top, looking south on Interurban Avenue to the former site of Tukwila station at 56th South in 1990 (All color photographs by the Author)

In this scene an interurban train can be seen southbound on 1st South just below Railroad Way. The passenger car storage area at the top of the photo is now occupied by Seattle's famous Kingdome. (courtesy of Wallace Swanson)

ALONG THE RIGHT OF WAY

The Airport Way viaduct was built in 1927 to separate streetcar and auto traffic from trains. PSE trains continued under the viaduct until the end of service; the rails were removed in 1930. Just below, the Northern Pacific siding into the Georgetown Steam Plant merged with one from the Interurban. The PSE siding also serviced the company carbarn. Tracks 1 and 2 are NP while 3 and 4 are Milwaukee and Pacific Coast. In the distance is the Georgetown brewery and OWRR&N Georgetown depot. (Both: NP joint facility photos from Daniel Cozine) Lastly, looking north on Duwamish Avenue (Airport Way) at Military Road. The Denny Renton clay plant at left center had both NP and PSE sidings into the plant. In the distance is the Seattle city limits sign and beyond, the PSE Chicago Avenue passenger shelter. (Asahel Curtis, Washington State Historical Society)

B
A
P
9
8
2
10
11

68

C
D
EX
B
3
A
G

1
2
3

Opposite left above, looking down Duwamish Avenue, south of Georgetown at the Northern Pacific main line on the left and the PSE tracks on the right. The tall stack of the clay plant is near the present day Boeing field terminal. This photograph graphically illustrates the interurban's problem: most of its length was paralleled by several main line railroads. The two photos at left are views of Renton Junction, looking south and east respectively. The high trestles in both views are the PSE's branch to Renton. "1" and "2" are the NP main and "3" is the NP's line to Renton and Sumas. "F" and "G" would be today's Grady Way's four lane viaduct over the tracks. To take these scenes today one would have to stand along the westbound lanes of I-405. "3" in the picture just left, would be the site of Longacres racetrack. (All photos: NP Joint Facility, courtesy of Dan Cozine) Above, a few views of Tukwila, looking south. The photo directly above was taken in either 1907 or 1908 when the line was still single tracked to Renton Junction. In the distance is the swing bridge over the Green River and left, the Nelsen barn. Just to the right of the bridge is a long stairway climbing Tukwila Hill. At one time there was a station here called "Whites", later renamed Black River after the line was double tracked. (Both Author's collection)

The photo at the top of the page shows the PSE trestle alongside the new Pacific Highway bridge over the NP and UP tracks at Reservation in Tacoma. At far left are the tracks into Bay Street yards that PSE shared with the TR&P. Above, in this photo, taken from the new highway bridge, PSE's abandoned trestle to its freight yards on the Tacoma tide flats, can be seen. Both pictures were taken in August of 1927. (NP Joint Facility Photos, courtesy of Dan Cozine)

Looking east on the newly opened Pacific Highway from the bridge over the Puyallup River. The line of poles on the right mark the right of way of the abandoned Puget Sound Electric. (Washington State Historical Society)

1st Avenue and Spokane Street looking north around 1935. PSE trains followed South Park cars as far as the trestle over the railroad tracks south of Spokane Street. The interurban left the South Park line at the top of the trestle and descended to Argo and Georgetown while South Park cars headed toward East Marginal Way. The trestle was eventually replaced by a concrete and steel structure with rails that were never used. (Author's collection)

I Remember The Interurban

By Clinton H. Betz

Today, after almost 70 years, what I remember of the Puget Sound Electric Railway are mainly bits and pieces. In some instances however, these memories are still as vivid as if they occurred yesterday. My home, in those days, was Renton. We were fortunate in town to have two railway lines into Seattle. You went either by the Puget Sound Electric interuban or the Renton & Southern, known as the Rainier Valley Lines. Early on, the Renton branch of the PSE was served by a shuttle car from the Seattle-Tacoma main line at Renton Junction. As I recall, outbound from Seattle you boarded the Tacoma local. The Renton shuttle was waiting, ready to board at the Junction. However, inbound was another situation as you generally had to wait for the Seattle bound train. A great number of Renton people preferred the Rainier Valley Line, because they did not have to transfer at the Junction. It was a time consuming nuisance, especially in winter.

I remember traveling with my parents to Seattle and going through the transfer in both directions. My dad disliked the transfers and I recall him talking to a man in Renton's business district about transferring cars. This man told my Dad the PSE shuttle car no longer existed, and the Renton car went directly through to Seattle.

In 1916 my father purchased a German built car called a Metz. On Sundays and summer evenings we'd ride on one of the few paved roads, the West Valley Highway which was paved with brick as far as Kent. Auto speed at that time was 15 miles an hour. Tires were poor with flats a common occurrence. In some places, the PSE paralleled the road. It was always a thrill to see a 2 or 3 car PSE train speed by at about 60 mph. From the turn at Renton Junction to beyond Auburn, the track was almost straight for 12 miles. This is where the PSE made up their time.

My Dad was a carpenter and worked at Camp Lewis when the base was first built during World War I. Most of the time, he came home on Saturday afternoons. I was about 8 years old at the time and with Mother's permission, I would go to the interurban depot and wait for Dad to arrive, generally on a 3 car train. He always had something for me. I was an only child and when Dad was home it was always a happy time. Mother was busy getting Dad's clean clothes and other items ready and on Sunday, after a special dinner, we left the house and walked the half mile to the interurban depot. He always left on the 3 PM train back to Camp Lewis. I remember standing across the street waiting for the train to pull out. We waved goodbye and on the walk home my thoughts would drift to next week.

Dad always had stories to tell. Once, he told of a ride down the valley, past Kent, with a bunch of loud drunken soldiers on board. As the conductor called out the stations of "O'Brien" and "Orillia", they mimicked, shouted and laughed, "this must be an Irish layout, singing out "O'Brien" and O'Reilly".

The PSE trains backed into the Renton depot from the wye west of Main Street between 5th and 6th Avenue (today, where Grady leaves Main Street). The wye was a tight turn, looping back to the incoming track and could accommodate four cars. The depot was located in one corner of the Renton Feed Company, occupying about one-half block on Main between Walla Walla Avenue (now Houser Way) and 4th Avenue. A family acquaintance was the Renton ticket agent and she generally opened the window about 10 minutes before departure time. The double tracks of the Pacific Coast Railroad ran east and west through Renton on Walla Walla Avenue, between the PSE depot and the business district. The tracks were also used by the Chicago, Milwaukee and St. Paul Railroad. I remember some frustrating episodes when a long Milwaukee freight train, moving slowly across Main Street, would prevent the interurban from pulling out on schedule. If you were alert, you could catch the nearby Rainier Valley car which left at the same time.

As a youngster one of my pals was the son of one of the owners of the Feed Company. We'd visit the wye which had a grease shack in the center, containing a barrel of grease and a mop stick. The sharp turn required track workers to apply grease from time to time to the inside flange of the rail; however, we helped too by liberally greasing not only the flange but the top of the rail as well. We watched from a distance as the cars almost stalled. It wasn't long before a lock was placed on the shack.

Another of our favorite places was where trains converted from trolley wire to 3rd rail. Generally the brakeman would hang out the train door on a belt, putting up or pulling down the trolley pole.

Many times, I watched the cars back up Main Street to the depot, the motorman hanging out the half door, the bell ringing. I recall motor car No.518 and 559 as regulars on the Renton Line. In those days, trains had one or two trailers.

After the 1st World War my parents built two greenhouses and opened a florist business, the only one in the little town of Renton. (Back then the population was about 3,000). My folks specialized in growing carnations. As Dad was a perfectionist, the blooms were top quality and in great demand with Seattle florists. The leading florist there always took everything available and the blooms were shipped several times a week via Renton Auto Freight. Once a week the florist would call for extras. That was my job after school, to take the flowers via interurban or Rainer Valley Lines into town. This task usually lasted through the growing season from October through May. I much preferred the interurban, since it seemed more train-like and rode better. It was quite a walk from the depot to 3rd and Madison where the florist was located. The way back was easier, all down hill, along 2nd Avenue. If I had a minute to spare, I always stopped by the Alaska Steamship Co. ticket office, located near Cherry St. In the window was the sailing schedule, with the names of the ships and their ports of call. I took the 4:05 PM car into Seattle and returned either on the 5:25 PM, a three car train, or the 5:35 PM which was a single car. I remember during the Puyallup Fair the PSE transferred the 600 series, or Jewett cars, to the Puyallup-Tacoma line. During this period, PSE had most of their equipment in service; substituting on the Renton branch were No.518, 558 and 559 and trailers.The Jewett cars were slow and lacked power. One evening in particular, while returning home from Seattle on No.558 the 5:35 PM train, the regulars began kidding Mr. Rummery, the conductor, remarking "Horace, where did you get this one"? This trip was a congenial and happy one. Going out on First Avenue South, the bell on car No.558 reverberated off the buildings and the high backed comfortable leather seats made the trip seem special. On most of my trips into Seattle I took the 4:05 PM, a single car which at times carried a crew of four. On leaving Renton, the car stopped where the PSE crossed over the NP tracks into the Renton Coal Mines. The brakeman ran ahead of the car to signal "all clear". This was repeated at Argo in Georgetown where the PSE crossed over the UP tracks. Going in on First South, the car backed into the PSE barn at Massachusetts Street and coupled on two trailers for the 5:05 return trip.

Mother and I made a trip to the Puyallup Fair in the early 1920's boarding the Tacoma local at the Junction. We were on car No.510, a heavy interurban. I remember the interiors, the stained glass, etched mahogany woodwork, the ceilings of dark green linoleum like material. It was a fast, smooth ride. We changed at Willow Junction to the Puyallup line, where we boarded the familiar Jewett cars. I remember crossing the interurban bridge into Puyallup, but remember nothing of the fair itself. On the return trip, we waited at Willow Junction for the Seattle bound train. Shortly, the three car train came roaring up and we climbed aboard. I ended up in the first car, No.512, and Mother in the middle car, an old trailer that got jerked about when the train started up. It was dark and the lights in the car flashed as the train crossed roads where there was a break in the third rail. The motors were whining away—a boring trip. Finally, we boarded the Renton local and reached home after a 1/2 mile walk from the depot. It was the end of a long, long day.

My folks visited with friends who lived on Earlington Hill. From there the railroad tracks up the valley were visible for a long distance. On a rainy evening, it was an interesting sight watching the interurban traveling up or down the valley. You could see the headlight as it made the curve at Auburn. At the crossings, the shoe on the car hit the wet 3rd rail, causing flashes that resembled 4th of July sparklers.

From 1922 to 1927, I rode either the PSE or Rainier Valley Line a great number of times. If I had to make a trip on Saturday morning, I would almost invariably walk to the waterfront, since next to the interurban, ships were my first love. Most passenger carrying vessels were concentrated at Pier 2, the Colman Dock, Grand Trunk Pacific and Pier 3.

Next door to the building, housing the Seattle depot, was Chauncey Wrights Restaurant which in its day was a well known upscale eatery. Huge steaks and seafood were displayed in windows to entice passersby. At the time the Tacoma train left before the Renton train. I enjoyed watching the Tacoma bound train leave. The boarding steps would come in and the conductor would give the high sign and off they went.

In 1927, while a sophmore in Renton High School, three fellow students and I made excuses to attend the Puyallup Fair. By this time, the Tacoma interurban consisted of one large car. We boarded the car at Renton Junction, where it stopped for passengers for Kent, Auburn and Tacoma. We went into a "hole" beyond Pacific City, before going up grade to cross over the old Pacific Highway. After the Seattle bound car passed we backed out onto the main line. The route from the tunnel to Edgewood can best be described as nine miles of twisting track which resulted in a very uncomfortable ride. By this time, the Puyallup branch was long gone; we had to go into Tacoma PSE depot to catch a bus to the

Fairgrounds. By this time, PSE roadbed was pretty rough, as business had dropped off, with apparently little road maintenance done. The return trip after dark seemed to accentuate the rough ride, the car lights blinked as we crossed roads and the howl of the motors was quite evident, a most boring adventure. This was my last experience riding on the Tacoma line. Little did we know at the time the closing of PSE was only 15 months away. I did continue to ride the Renton-Seattle line several more times.

In 1926, my father purchased a new Chevrolet sedan. I had driven the old Metz since I was 13 years old. The following year I started to drive into downtown Seattle after school with the flowers. The first wholesale florist in Seattle opened for business in the basement of the Seaboard Building at 4th and Pike Street. On the return, I brought back florist supplies and flowers we didn't grow to satisfy the demand of our growing business.

I remember on January 1, 1929, Mother and I were invited to Bremerton for a New Year's Day dinner. We had a 1928 Chevrolet by this time and I had my first experience driving a car on the autoferry. At the time newspapers carried the account of the closing of the Puget Sound Electric RR as of midnight, December 30, 1928.

During the summer of 1929, some of us fellows walked across the trestle at Renton Junction. It was high and had an "S" curve. On the turns, I still can recall the four or five ties spiked together and the inward slope of the track.

The arrival of the automobile brought better roads and the public flocked to the new convenience---THIS WAS PROGRESS! In my case, I was no longer confined to a rigid schedule, waiting for or hassling heavy packages on crowded cars. Like so many others I embraced the convenience of the automobile and in doing so, contributed to the demise of the interurban.

Car No.555 at Renton Junction. Without a train number showing, this was probably when the company was operating a shuttle service between the Junction and Renton. Sam Neal, the Renton Junction operator, is on the right. Neal became an operator on the Pacific Coast Railway after the interurban quit. The identity of the crew not known. Note the penny scale on the right. (Author's collection)

Above left, car 523 in use as a tool shed by Tacoma Municipal Belt. It was later sold, moved to Federal Way and became a home. In the 1960's it was bought and moved again to the trolley museum at Glenwood, Oregon. (Vernon Rutledge) Above, former PSE No.504 as a logging coach for Weyerhaeuser at Vail, Washington in 1938. (Harold Hill) At left, No.556 was a diner in Kent until 1960. Car No.514, below, was a companion to the 556 and was part of the same restaurant complex. In the 1960's it was moved a short distance to become part of another building. It was later destroyed by fire. (Both: Lawton Gowey)

Left, one of the Renton Line's Jewett cars is shown stored at TR&P's Bay Street yards in Tacoma, circa 1935. For some reason this car and several others of its class were stored there for years after the interurban quit. One 600 became a restaurant on Highway 99 just up the hill from the Duwamish bridge. Below are several views taken by Harold Hill in 1938 of former PSE cars on Washington state logging roads . All appear the worse for wear. Dempsey Logging Company No.700, ex-PSE No.513, is seen at Lake Kapowsin. Ex-517 as Saginaw Logging No.26 is at Brooklyn, Washington. Below, former PSE 558 as a crew car for St. Paul-Tacoma Logging Company. (All photos, Harold Hill)

Following the demise of the Puget Sound Electric, several cars were sold for use on Seattle City Light's tourist operation out of Rockport, Washington. In the view above four former PSE cars are in tow behind City Light's locomotive No.6, en route up the Skagit River above Rockport. (Seattle City Light, courtesy of the late Ted Carlson)

City Light refurbished the former PSE cars they bought in 1929 and kept them in excellent condition until the end of operations in the early 1950's. No 9, left, is the former PSE No.515. Below, ex-PSE No.527 was not as fortunate as the former Oregon Electric cars, like the No.11, that went south to become part of Andy's Diner in Seattle. The photo of the burned 527 was taken at Newhalem in the late 1950's. (All, Harold Hill)

Riding The Interurban

We're waiting to board the interurban to Tacoma. It's a little after 10:30 in the morning of a cool, partly cloudy fall day in 1928; normal for Seattle this time of year. Earlier it had been foggy down through the valley making some early morning trains a few minutes late. But the fog has lifted now and we've been told our train will be on time.

We're standing in front of the interurban station on Occidental Avenue, between Yesler Way and Washington Street. The interurban will come into town on First Avenue over the tracks of the Seattle Municipal Street Railway to Yesler Way. There, it will turn east for a block, then south on Occidental.

PSE's telegraph, ticket office and waiting room are located on the ground floor of this huge four story, red brick building, which was in the center of Seattle's business district when the Interurban began in 1902.

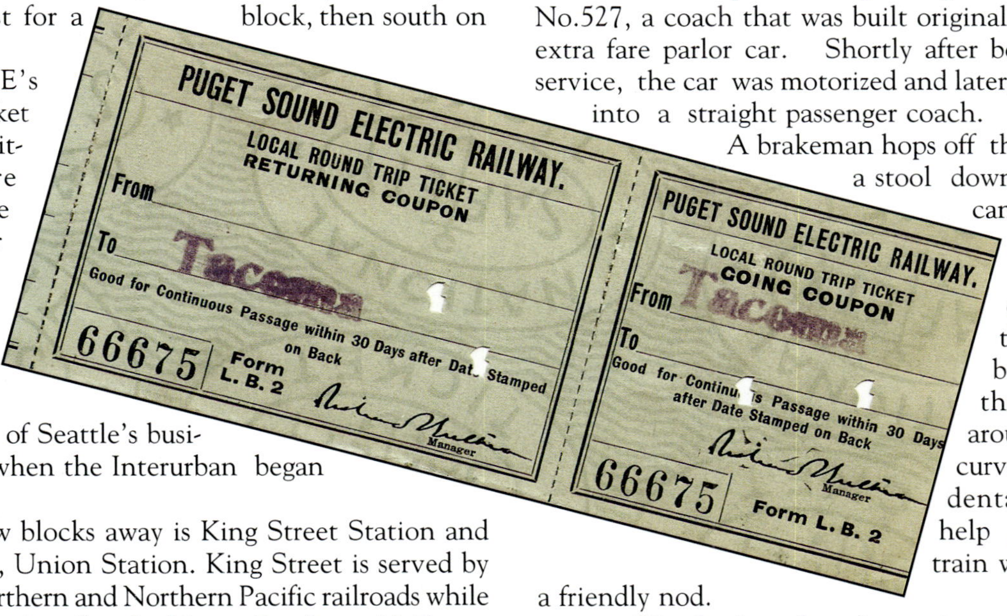

A few blocks away is King Street Station and further down, Union Station. King Street is served by the Great Northern and Northern Pacific railroads while Union Station handles the Union Pacific and Milwaukee Road. Further west is Seattle's busy waterfront on Puget Sound where ships sail to various ports around the world. From famed Colman Dock ferries and passenger steamers travel to all ports along the Sound. Colman is also where two famous ships, "Tacoma" and "Indianapolis", make the run to Tacoma every two hours. Both steamers were the Interurban's chief rival until auto stages began some eight or nine years ago.

The Interurban is already in the hands of the receiver and there's been talk of abandonment. The Interurban has come under criticism recently for its dangerous third rail as well as several bad accidents. Puget Sound Power & Light, the lines parent company, is already running buses on highways paralleling the interurban. And they have just released a survey to determine what bus schedules would best serve the interurban's patrons. So it would seem the end is near.

We hear the interurban approaching on Yesler and there is a general hustle and bustle as we prepare to board. A handsome two car train, painted dark green with gray roof and gold lettering, rounds the corner and glides to a stop in front of the depot. In busier times, PSE ran longer trains, sometimes with three or more cars, but now two cars will suffice for hourly service and single cars on the half hour.

Our train consists of motor No.514, a combination express and passenger car, and trailer No.527, a coach that was built originally as a deluxe extra fare parlor car. Shortly after being placed in service, the car was motorized and later, it was rebuilt into a straight passenger coach.

A brakeman hops off the 527 and sets a stool down so passengers can reach the car steps. Meanwhile, the conductor on the 514, who'd been guiding the trolley pole around the sharp curve onto Occidental, begins to help people off the train with a smile or a friendly nod.

We climb on board as the conductor steps into the telegraph office for his orders. Although we're going all the way to Tacoma, we take a seat in the lead car, which is usually reserved for local passengers. Our car, the 514, is composed of baggage-express, smoking, and coach compartments, with a swinging glass door between the smoker and the coach sections. Fares are collected and rung up on huge registers at the front end of the car.

This train is No.45, which leaves Seattle daily at 11:00 AM. It's a local, making all stops between Renton Junction and Tacoma. It also stops at stations between Georgetown and the Junction for passengers bound for points south.

Our conductor has returned from the telegraph

office with the orders and walks to the head end where he compares watches with the motorman. Hurrying to the rear of No.514, he picks up the stool and hollers, "All Aboard".

Once aboard he gives a two-bell signal telling the motorman its okay to proceed. The motorman rings the floor gong, releases the air and opens the controller a notch and we slowly drift away from the depot. Past Washington Street we roll along to Jackson. Here, a "fare" boards on the run as we round the curve and proceed to 1st Avenue.

There's a momentary stop for a South Park streetcar that has arrived at the north side of the intersection. Finally, we move out ahead, for if we get stuck behind the city car it would mean we'd be late all the way to Tacoma.

We now cross King Street and slow to let a Kinnear trolley get out of our way; it had been using the crossover in the middle of the block for its downtown turnaround.

At Railroad Way, we glide to a stop for the steam road crossing. A watchman steps out of his little shanty and flags us across the tracks. From this point on, the motorman will use the overhead locomotive type air bell and when we reach Georgetown, he'll use the air whistle. Inside city limits the whistle is used only in emergencies.

At Massachusetts Street, we slow for a turnout leading west where the PSE freight shed and substation house are located. At one time a carbarn was located here but has since been torn down. Several cars from the Renton line are stored here and occasionally, a car from the main line.

Our train grinds over the cobblestones, passing the big Sears Roebuck mail order store. At Hinds Street, we slow for a West Queen Anne car. Arriving at Spokane Street we stop for passengers who might have walked over from the city's elevated line on the waterfront, several blocks west. This is where streetcars from the Alki, Fauntleroy, West Seattle and Lake Burien lines travel into downtown Seattle.

Just beyond Spokane Street we stop for another steam road crossing; the Northern Pacific's line to Harbor Island and West Seattle that serves several sawmills, flour mills, warehouses and steamship piers. We are on the South Park line now and after our train crosses the tracks, we begin to climb a wooden trestle. At the top we switch off to another trestle before descending to the ground. Originally, this trestle was over the tide flats, but in time they were filled in and a railroad switching yard built there. Now we race along between the Union Pacific and Northern Pacific tracks for a quick run to Argo crossing.

We pass a semaphore signal showing clear, and enter the single track, passing under the new Argo Bridge. Crossing the UP tracks we swing to the left, entering double track on the city's South Seattle line.

Before Argo overpass was opened the crossing was a nightmare for trolley crews, main line railroads and motorists. The South Seattle cars crossed the NP and Milwaukee tracks on single track then joined the Interurban to cross the UP on double track.

We roll up Duwamish Avenue on the east side of the brick paved street, passing a huge brewery on our left. The South Seattle line splits off here and we continue on to Georgetown where a stop is made to pick up a few passengers.

Rounding a curve, we head south between the NP tracks and the highway. Three blocks south of Georgetown station we slow for a switch into Seattle Muni's carbarn and shops. A high board fence runs along the entire length of the yard to a concrete building. Over the top of this fence we can see the roofs of several streetcars. At the south end of the carbarn is another turnout. This is used by city cars and PSE to switch carloads of coal into the PSP&L's Georgetown steam plant, located just west of the carbarn. This steam plant is one of the PSE's main sources of power.

Now our whistle sounds for Gorgiats, Van Asselts and Chicago Avenue. Rolling past the new King County Airport, our motorman keeps a sharp eye on oncoming automobile traffic. The highway here is practically up against the southbound track, and occasionally our motorman has to use his whistle to warn careless motorists. A new 1929 Model A Ford is doing its best to keep up but we pick up speed and soon lose it.

Beyond the airport the whistle sounds for yet another steam road crossing, the NP siding into the Denny-Renton brick plant. PSE also has a siding here. A swinging gate with a red lantern is across the steam track, protecting the interurban.

Soon, the highway turns west and we are on our own. After a slight turn we slow for the Meadows wye. At one time, a popular race track was located here and in its heyday, the Interurban did land office business hauling race fans. In years past a telegraph operator was stationed here, but now there's just a telephone and the wye is used only for emergencies.

Beyond Meadows we approach Southside, where our train makes the change from overhead trolley to third rail. The conductor has opened the rear vestibule door of the 514 and awaits the signal from the motorman that we've reached third rail territory. As soon as the whistle sounds, he pulls down the pole, hooks it on the roof and steps inside, giving a five bell signal to the motorman, who acknowledges with two short blasts from his whistle.

Shortly, we pass our first block signal. The semaphores on PSE are numbered according to their distance from Tacoma; this one, No.299, is 29.9 miles north of Tacoma. These signals were installed back in 1914 between Southside and Bay Street in Tacoma.

The motorman whistles for Cardmoores Station but we don't stop. Like most of the PSE's smaller stations, this is just a little covered platform used as a shelter during inclement weather. No passengers are waiting today as the little wooden semaphore used to signal the motorman, is in the "down" position. If it were raised, signifying a waiting passenger, the motorman would answer with two shorts blasts on the whistle.

We round a curve and start up a slight grade. The whistle sounds for Duwamish Station with one long blast and then another for the Duwamish bridge, a single track, steel span that replaced an earlier wooden drawbridge.

The motorman slows our train to 15mph to cross the bridge. Once across, we enter double track and racing around a curve on a trestle, pass a high embankment before going downhill to Quarry Station. A rock quarry is located here with a siding. PSE has hauled many a carload of rock from this quarry and at one time, the buildings here were lighted by power from the interurban.

I remember my father had a team of horses get away from him at the quarry in 1909. Both horses were electrocuted as they crossed the third rail and trains were held up pending removal of the bodies. Through the years many animals have lost their lives due to the third rail, and every PSE car carries a long pole which is used by train crews to lift small animals whenever they were unfortunate enough to tangle with it. Larger animals required power to be turned off temporarily. The third rail is out in the open, without any protective cover and needless to say, crews and regular passengers have a healthy respect for it.

We're alongside the Duwamish River now and running downgrade toward Allentown station. Beyond Allentown we slow for the Riverton road crossing which is on a curve with a ten mile speed limit. The crossing is protected by a wigwagging bell danger sign. The whistle blows as we cross the intersection for Riverton, just beyond the crossing.

The passenger signal board was up and we have

made the stop. The station is on our left and a freight shed is to the right. On the platform are several milk cans waiting for the PSE milk train. Our passenger climbs on board and we take off.

Speeding along on a stretch of straight track for about a mile, we pass Mortimer station where the highway again parallels the tracks. Around the next curve we fly past Foster and the new Foster golf links. Beyond Foster is a signal showing green and a crossover.

We take another curve at Tukwila and our motorman whistles for Black River. Signal 259 burns green but we slow for single track and a curve to the left. We skirt the Green River (the name changed at Black River) and pull up onto the trestle approach to the drawbridge over the river.

Once across, we ease to a stop at Renton Junction. Renton line trains branch off here on a high trestle over steam railroad tracks. After the brief stop our train proceeds south, finally leaving the trestle for solid ground.

Renton trains once operated on an hourly schedule, but because of low patronage, service has been reduced to morning and evening service. In the middle of the day a shuttle operates hourly between Black River and Seattle. Renton is served not only by a new bus line but also the Seattle & Rainier Valley Railway. Renton is also the site of a large, company owned coal mine and many carloads of coal are hauled nightly into Seattle. This is one of PSE's larger sources of revenue.

South of Renton Junction the track straightens for twelve miles and the motorman opens the controller a few notches for high speeds are the rule here. Quite often the interurban would race trains of the Milwaukee Road on this stretch, since the two lines parallel each other from Renton Junction to a point south of Auburn.

The conductor gives a three bell signal that a passenger will be getting off at Orillia, the next station. The motorman gives his answer with two shorts on the whistle. We stop briefly, then two pulls of the bell cord and we head out again.

Orillia features a seven-car capacity passing track, where trains meet frequently. Beyond the siding we cross a small stream on a wooden trestle. Then the whistle blows for Cochrane, but no stop is made. Next is O'Brien and the passenger signal board is up. We are exactly a half hour out of Seattle and right on time. At O'Brien is another passing track (which is used very seldom) and a siding. Two bells from the conductor, and we start again, making a run to Kent.

The motorman whistles for the trolley pole to be raised as we approach the Kent city limits; where we leave the third rail and use overhead trolley through the city. This time, the brakeman opens the rear vestibule door and as soon as we come under the wire he raises the pole.

Our big overhead bell is ringing steadily as we cross a busy thoroughfare and gradually come to a stop at Kent station, a two story wooden building with a covered porch around three sides. This station houses the PSE offices, a bunk room for train employees, as well as the dispatcher's office and waiting room.

New orders are received and we start again. A passing track is on our left as well as the Kent substation, and beyond, is the Kent carbarn and car shops. We pass a siding on our right and four yard tracks on our left. Four additional tracks lead into the carbarn, while just beyond the yard is a loop track.

Two blocks south of the Kent station we pick up the third rail again. At Willis Street crossing, south of Kent, we pass the scene of a nasty head-on collision that occurred several years before.

A few miles south of Kent we cross the Green River for the last time and enter double track before pausing at Thomas station. We'll be on double track almost to Auburn. Somewhere along this stretch we're scheduled to pass a northbound local, which left Tacoma the same time we left Seattle.

Our whistle sounds for White River, and the station and road crossing at Meredith. At Christopher, we make a stop for a passenger and while there, the northbound local goes by. At the outskirts of Auburn, we leave the third rail again to pass through town. Entering single track, we cross a busy street, and stop at Auburn station. This is the largest town on the line and, like Kent, is a stop for limiteds as well as locals.

With new orders, we leave Auburn as the motorman whistles two longs and two shorts for a grade crossing. Now we swing to the southwest and pass Farrow siding. Next is Algona and Pacific City, where stops are made for passengers.

We start upgrade now, slowing for Bluff's siding. Often trains meet here when a southbound is late for the scheduled meet at Edgewood. Crews dislike this siding because it means backing out and then taking a slow climb uphill to Edgewood.

No stop is made at Bluffs today. We cross over a brick highway on a steel span and take a sharp curve to the left. The highway, until this year, was known as the Pacific Highway but when a four lane road was opened west of here, it changed to West Valley Highway.

Uphill we travel, taking one curve after another. Soon we pass a green signal and round a long curve that goes through a cut. Here we enter the only tunnel on the line, a 182 foot long wood lined affair. Once through, our track straightens for a short distance, then swings left, past signal 103. The going here is slow as we climb steadily until Edgewood is reached.

Posted speed through this area is 20 mph for passenger trains, 15 mph for freights. Eight minutes is the least amount of time allowed for the four and a half miles between Bluffs and Milton. This run is scheduled for six minutes for the 2.22 miles to Edgewood and four for the 2.27 mile downhill run.

We round a curve to the left, pass an abandoned logging spur as the whistle sounds for Jovita and its grade crossing. A signal is up and we make a stop. Passengers board the 514, and with two pulls of the bell cord, we are on our way.

Just beyond Jovita we pass a sawmill with its own spur. In early years, the interurban hauled logs to the mill and finished lumber to the cities. At one time PSE actually owned hundreds of acres of virgin forest in southern King County.

Shortly the track levels out as we pass signal 91 and see PSE's gravel pit off to the right.

We're nearing the Edgewood passing track. The motorman has sounded two longs and one short, indicating a meet. Our conductor responds with six pulls on the bell cord. The brakeman has come forward and as soon as we stop, he'll jump down to open the switch so we can move into the siding.

After entering the siding, our brakeman closes the switch and locks it, and jumps aboard the car as we move up to the far end of the siding to wait for the limited, which left Tacoma at eleven o'clock.

Several minutes pass before we hear the limited grinding its way up hill. Finally, it comes into view, its motorman whistling for the station and the road crossing. With a high sign from each crew, the limited rushes by without stopping.

Our conductor signals the motorman and we move through the spring switch, out of the siding and cross the road to stop at the station. Two bells and we're ready to leave; just beyond the station the motorman tests the air brakes in accordance with Rule 14. Air brakes are tested northbound and southbound on the tangent approaching the curve at the north end of Edgewood passing track.

We take a curve to the left, pass through a small cut and begin our descent. Passing the site of the old Peerless Coke Works siding, we swing to the right through a deep cut. Around still another curve and another cut we pass signal 69, which is showing green, allowing us to proceed.

We are riding high above a fast moving stream, heading south toward Milton. We slow for the Milton passing track and, keeping to the main, we pass the Milton substation. Our train passes through an area called Dead Man's Cut and pulls up abruptly at Milton depot.

Passengers board here and, after we have come down out of the hills, we straighten out for a westward run through the Puyallup Valley to Tacoma.

A signal board is spotted at Fife and we slow for the station. Then the whistle sounds for Gardenville, but we pass without stopping. Signal 47 whizzes by and then we slow to 15 mph for Sicade passing track. This was formerly called Willow Junction and was where the Puyallup Short Line turned off to the southeast.

The station of Goldau and Tidehaven fly by, and soon we start up a small rise to cross the Puyallup River on a steel span.

Leaving the bridge, we remain on a high trestle, which crosses the Puyallup yards and Portland Avenue. We come to a stop at Bay Street where we leave the third rail for the last time. A sharp curve to the right and a fairly steep descent to Puyallup Avenue is next. Our motorman gets two bells and we move out onto the avenue behind a Portland Avenue streetcar. Moving up Puyallup Avenue, we make a stop for the steam road crossing, and then slow for the entrance to the McKinley Avenue carline. A few blocks further we stop to allow passengers to leave the train before proceeding up Pacific Avenue.

The Portland Avenue Birney stops at the Union Depot and we do too. Just beyond the depot, a slow freight is passing, so the motormen of both the Birney and our interurban wait until it clears.

At Thirteenth Street we stop for the cable line; it has rights over the Interurban at this intersection, but at 11th, the interurban has the right of way. At Ninth, we turn right one block and then a left turn takes us to the depot on A Street. Our conductor yells, "Tacoma" and picks up his little stool and steps off as the train makes its final stop.

Roster of Equipment

CAR NO.	TYPE	BUILDER	DATE	SEATS	LENGTH	MOTOR	TRUCKS
500	Motor Combo	Brill	1902	36	42'6"	4 GE 66	Brill 27E1-1/2
501	Motor Coach	"	"	42	"	"	"
502	Motor Combo	"	"	36	"	"	"
503	Motor Coach	"	"	42	"	"	"
504	Motor Combo	"	"	36	"	"	"
II 504	" "	TR&P	1914	64	58'	4 GE 66b	Baldwin 78-35
505	Trail Coach	Brill	1902	40	42'6"	----	Brill 27E1
506	Motor Combo	"	"	36	"	4 GE 66	Brill 27E1.5
507	Trail Coach	"	"	40	"	----	"
508	Motor Combo	"	"	42	"	4 GE 66	"
509	Trail Coach	"	"	40	" -	----	Brill 27E
510	Motor Combo	St. Louis	1907	58	55'	4 GE 66	Baldwin 78-30
511	Trail Coach	Stevenson	1903	40	42'6"	----	Brill 27E
512	Motor Combo	St. Louis	1907	58	55'	4 GE 66	Baldwin 78-30
513	Trail Coach	Stevenson	1903	40	42'6"	----	Brill 27E.5
514	Motor Combo	Cin.	1910	58	55'9"	4 GE 66	Baldwin 78-30
515	Trail Coach	Stevenson	1903	40	42'6"	----	Brill 27E1.5
516	Motor Combo	P S E	1912	58	55'	4 GE 66	Baldwin 78-30
517	Trail Coach	Stevenson	1903	40	42'6"	- ----	Brill 27E1.5
518	Motor Combo	St. Louis	1909	46	50'	4 GE 66B	"
519	Trail Coach	Stevenson	1903	40	42'6"	----	"
520	Motor Combo	St. Louis	1909	46	50'	4 GE 205	"
521	Trail Coach	"	1907	58	58'	----	"
523	Motor Coach	"	"	58	58'	4 GE 66	Baldwin 78-30
525	"	"	1910	58	58'	2 GE 66	See right
527	"	"	"	58	58'	"	"
529	"	Cin.	"	58	58'6"	"	"
550	"	St. Louis	1905	48	41'	4 GE 90	Brill 27G1
551	"	"	"	44	41'	"	"
552	Trail Coach	"	"	48	41'	----	"
553	Motor Coach	"	"	44	41'	4 GE 90	"
554	"	"	1906	44	41'	"	"
555	"	"	1903	48	41'	"	"
556	"	"	1909	56	50'	"	Baldwin 78-30
557	"	"	1906	56	41'	"	Brill 27G1
558	"	"	1909	60	50'	"	"
II 558	"	"	"	60	50'	4 GE 66	Brill 27E1.5
559	"	"	"	60	50'	4 GE 205	"
560	"	"	"	62	50'	2 GE 66	"
561	"	"	"	60	50'	4 GE 66	"
600	"	Jewett	1907	56	50'4"	4 GE 80	Brill 27 MCB
601	"	"	"	"	"	"	Baldwin 56-22
602	"	"	"	"	"	"	"
603	"	"	"	"	"	"	"
604	"	"	"	"	"	"	Brill 27 MCB
605	"	"	"	"	"	"	"
606	"	"	"	"	"	"	"
607	"	"	"	"	"	"	"
608	"	"	"	"	"	"	"
1000	Business Car	St. Louis	1906	37	49'0"	4 GE 90	Baldwin 72-25
1600	Box Motor	P S E	1903	--	40'0"	4 GE 66	Baldwin 78-30
1601	"	"	"	--	"	"	"
1625	Locomotive	"	"	--	38'0"	"	"
1626	"	"	"	--	"	"	"
II 1626	"	"	1914	--	40'0"	4 GE 66B	Baldwin 74-30
1627	"	"	1907	--	"	4 GE ;66	"
1650	Wrecker	TR&P	1910	--	42'6"	4 GE 57	Baldwin 78-30

DISPOSITION/DATE	REMARKS
Dismantled 1916	Electrical equipment to 516, body to II 504
Dismantled 1916	Electrical equipment, trucks to 558
Burned 1916	No information
Sold 1929	Skagit #10
Rebuilt 1914	Body to #504
Sold 1929	Weyerhaeuser #718, Vail, WA
Dismantled 1913	No information
Dismantled 1914	Electrical equipment to #504
Sold 1929	Skagit #12
	No information
Unknown	Out of service 1919
Burned 1927	No Information
Unknown	In collision, Riverton; repaired
Sold 1929	No information
"	Crew coach, Ohop, WA
"	Beer parlor, Kent, WA
"	Skagit #9
"	Rebuilt from parlor car #521
"	Crew coach, Brooklyn, WA
"	Ex-561, crew coach, Ohop, WA
	No information
Wrecked 1927	Ex-559; wrecked at Kent, WA
Rebuilt 1921	Became 516; orig. parlor-observ.
Sold 1929	Crew car; Tac. Municipal Belt Ry
Burned 1921?	Burned on main line
Sold 1929	Skagit #11
"	No information
Sold 1916	To TR&P Dec. 1916, #57
Sold 1912	To TR&P, Apt. 1912; #56
Sold 1917	To TR&P, #60
Sold 1911	To TR&P, #54
Sold 1912	To TR&P, #55
Sold 1917	To TR&P, #58
Sold 1929	Black River shuttle car
Sold 1917	To TR&P. #59
Rebuilt 1916	Motorized w/eqpt of 501 7/1916
Sold 1929	Crew coach, Ohop, WA See 558 above
Rebuilt 1921	See 520
Sold 1929	Rebuilt to parlor motor, 7/1916
Reuilt 1921	Rebuilt to 518 w/eqpt from 1601
Sold 1929	To TR&P, #146, scrapped 1932
"	To TR&P, #140, scrapped 1932
"	To TR&P, #134, sold 1932
"	To TR&P, #132, sold 1932
"	To TR&P, #138, scrapped 1938
"	To TR&P, #139, scrapped 1932
"	To TR&P, #145, scrapped 1932
"	To TR&P, #131, scrapped 1932
"	To TR&P, #141, scrapped 1932
Sold 1920	To PNT, #80, tavern, Everett, WA
Dismantled 1926	Renton Mine switcher
Dismantled 1921	Equipment to #518
Sold 1929	Skagit #80

NOTES

NOTE: There can be room for reasonable doubt wherever PSE is listed as builder; it would probably be more accurate to list TR&P as builder.

MOTOR H.P
66...125
66B..125
Cars 525, 527,
cars 529 had one motor truck (Baldwin 78-30) and one trail truck (Baldwin 72-30) for many years. Later other trucks were motorized also.

The 1600 Class originally was the 600 Class; upon the acquisition of passenger motors 600-608 from the Spanaway and American Lake lines of Pacific Traction, Tacoma, the digit "I" was prefixed to PSE's 600's.

All Skagit cars scrapped in 1954.

MOTOR HORSEPOWER

66.....125
66B...125
205... 100
90..... 50
80..... 38
57..... 60

Car histories compiled by Harold A. Hill, Lawton Gowey and Warren Wing from official PSE records and personal observations.

PUGET SOUND ELECTRIC RAILROAD INVENTORY AT TIME OF ABANDONMENT 1929

Trackage in Miles of Equivalant Single Track

MAIN LINE:						
In Gravel Ballast						
	29.013	11.681	1.377	1.282		43.353
On Bridges&Trestles						
	2.532	759				3291
IN GEORGETOWN:						
Paved:						
Brick paved,						
Duwamish Ave.	480					480
RENTON BRANCH:						
In GravelBallast						
	1,912	455		849		3.216
On Bridge& Trestles						
		771				771
IN RENTON, PAVED:						
Brick Paved,						
Main Street		58				58
Plank Paved, Main St.						
		152				152
KENT YARDS						1.067
TOTALS	34,438	12,920	1,832	2,131	1.067	52,388

The approximate tonnage of rail in the above listed pieces of track is as follows:

80 Lb. H. S.	60 Long Tons
70 Lb. A S C E	5,130 Long Tons
60 Lb. A S C E	460 Long Tons
56 Lb. A S C E	10 Long Tons
60 Lb. A S C E	460 Long Tons
56 Lb. A S C E	10 Long Tons
	5,6000 Long Tons

BRIDGES AND TRESTLES:

Description:		
Truss Span, Combination	Bay Street	139 Ft
" " Pony Howe	N.P. Ry.	92 "
" " Combination	Puyallup River	108 "
" " Combination	Puyallup River	108 "
Eye Beam	C M ST P & P RY	56 "
Plate Girder	Bluffs	53 "
Truss Span, Howe	White River	96 "
Eye Beam	White River	36 "
Eye Beam	Renton Junction	26 "
Truss Span, Howe	Green River	96 "
Truss Span, Steel	Duwamish River	208 "
Plate Girder	North Bound Track, Argo	84 "
Plate Girder	South Bound Track, Argo	100 "
Wooden Trestles	Main Line	16,172 "
Wooden Trestles	Renton Branch	4,073 "

TUNNELS:

One Tunnel, timber lined - length 182 feet, located near Bluffs

ROADWAY TOOLS

# Of Units	Description
1	Gasoline Section Car, # 16 Kalamazoo
1	Gasoline Section Car, # 25 Kalamazoo
5	Gasoline Section Car, 6 HP Fairmont
6	Push Cars
2	Velocipede Hand Cars, Sheffield
1	Electric Hoist with 3 HP D.C. Motor
5	Bridge Jacks
12	Track Jacks
3	Rail Benders

BUILDINGS:			**SQ. FT.**	**ITEM**
1	Pass. Station & Office Tacoma		5,602 sq. ft.	# 180
1	Pass. Station, Sicade		280 " "	16
1	Pass. Station, Auburn		1,640 " "	71
1	Pass. Station, Kent		2,820 " "	101
1	Pass. Station, Renton Jct.		572 " "	125
32	Pass. Stn Shelters, Open (located at various points along R/W)		1,280	
52	Station Platforms		36,000 sq. ft.	
1	Car Barn	Kent	18,776 " "	100
1	Blacksmith Shop, Kent		520 " "	100
1	Oil House	Kent	144 " "	100
1	Car CleanersTool House, Kent		84 sq. ft.	100
1	Roadway Div. Store House, Kent		900 sq. ft.	99
1	Roadway Div.Oil House	Kent	144	99
1	Roadway Div.Tool House	Kent	456	99
1	Bonders Tool House	Kent	512	99
1	Bonders Tool House	Sicade	120	16
1	Roadway Div.Tool House	Milton	120	26
1	Roadway Div.Tool House	Jovita	120	42
1	Roadway Div.Tool House	Auburn	800	71
1	Roadway Div. Tool House	W.Hill	120	126
1	Signal Dept. Shop Bldg16' x 20'-11" to plate shingle roof, drop siding, Unfinished inside			71

NOTE: Substation buildings are listed with Transmission & Distribution System.

PASSENGER CARS

No. of Units — Car Numbers & Description

1 # 504, Combination Passenger and Baggage Motor Car.Body length 58'0", equipped with four G.E. 66 Motors and Baldwin 78-35 trucks

3 #'s 512, 514 and 516 Combination Passenger and Baggage Motor Cars. Body length 55"0",equipped with four G. E. 66 motors and Baldwin 78-30 trucks

2 #'s 518 & 558 Combination Passenger and Baggage Motor Cars. Body length 50'0", equipped with four G. E. 66motors and Brill 27 E 1-1/2 trucks

1 # 520 Combination Passenger and Baggage Motor Cars. Body length 50'0", equipped four G E 66 motors and Baldwin 78-30 trucks. NOTE: This car was wrecked in September 1925 and has not been rebuilt

2 #'s 523 and 527 Coach Motor car. Body length 55'0" equipped with four G E 66 motors and Baldwin 78-30 trucks.

1 # 529 Coach Motor Car. Body length 58'6", equipped with four G E 66 motors and Baldwin 78-30 trucks.

1 # 556 Coach Motor Car. Body length 50'0", equipped with four G E 90 motors and Baldwin 72-25 trucks.

1 # 503 Coach Trailer. Body length 42'6", equipped with Baldwin 72-30 trucks

6 #'s 506, 507, 511, 513, 517 and 519 Coach Trailers Body length 42'6", equipped with Brill 27 E 1-1/2 trucks.

1 # 515 Coach Trailer. Body length 42'6", equipped with Brill 27 trucks

FREIGHT LOCOMOTIVES & SERVICE CARS

No. of Units — Car Numbers & Description

1 # 1626 Electric Locomotive, Center Cab type. Steel construction. Weight 45 tons. Length 40'0". Equipped with four G E 66 motors and Baldwin 74-30 trucks.

1 # 1627 Electric Locomotive, center cab type. Wood construction. Weight 45 tons. Length 40'0", equipped with four G E 66 motors and Brill 27 E 1-1/2 trucks

1 # 1625 Electric Locomotive, partially equipped. Center Cab type. Wood construction. Equipped with Brill 27 E 1-1/2 trucks without electrical equipment.

FREIGHT CARS AND MISC.

28 Flat Cars, 25 boxcars, 10 Gondolas, 3 Caboose cars. 7 Gasoline Section or Motor Cars, a number of Hand Push Cars.

1 Electrical equipment complete, including four GE 205 motors, type DB 141 contactors, DB 407 reverser, DB102 circuit breaker and type CG resistance.

7 Type GE 66 railway motors

3 Type GE 66 armatures

2 Type GE motor cases with fieds or armatures

2 Passenger car truck frames, Baldwin type 72-30

2 Air compressor, Westinghouse type D2

2 Compressor armatures, Westinghouse type D2

Plus misc. freight and passenger car axles with 5"X9" and 4-1/4"X8" journals

INDEX

Bold face indicates photographs

BIBLIOGRAPHY

Auburn Argus
Auburn Republican
Auburn Globe Republican
White River Valley
Historical Society Museum
Tacoma Ledger
Tacoma News Tribune
Puget Sound Electric Journal
Tacoma Public Library
Arthur D. Kempster Scrap Books
Special Collections,
University of Washington
Puget Sound Electric Journal
Special Collections,
University of Washington
Seattle Times
Seattle Star
Seattle Post Intelligencer
Seattle Public Library
Puget Sound Electric Railway
Interurbans Special No.23, Swett, 1960

INTERVIEWS

Frank Cummings
Lawton Gowey
R.A. Hansen
Roy Henderson
Harold Hill
Roy Kelly
Virgil Napier
C.S. Paine
Horace Rumery
Henning Sunby